7 PRINCIPLES

ULTIMATE

FOR WINNING IN THE AGE OF

PARTNER

ECOSYSTEMS, AI, AND CO-SELL

VINCE MENZIONE

Ultimate Partner
7 Principles for Winning in the Age of Ecosystems, AI, and Co-Sell

Published by **Ultimate Partner Press**

First Edition, 2026

Published in the United States

Ultimate Partner Press

Library of Congress number: 2026922394

ISBN: 979-8-952334-01-4 (*Hardback*)

ULTIMATE
PARTNER
PRESS

DISCLAIMER

This is a work of nonfiction. The events and experiences it describes are true, recounted as faithfully as the author's memory allows. Many of the quotations throughout the book are drawn directly from interviews conducted on the Ultimate Partner podcast and recorded conversations at UP LIVE events. These have been reproduced with care, without altering the speaker's meaning. The views expressed by those interviewed are their own. In a few limited instances, the author has changed the name of a company and altered certain identifying details. The opinions expressed are the author's, and nothing in this book is intended to disparage any company, product, or individual.

DEDICATION

To those just getting started, out to make their mark on the world.

This is the book I wish someone had handed me 40 years ago.

CONTENTS

FOREWORD

By Jay McBain, Chief Analyst–Channels, Partnerships & Ecosystems, Omdia

I have spent my career studying what separates the organizations that win in the technology ecosystem from those that stall, plateau, or disappear.

After conducting three decades of research, having thousands of conversations with channel chiefs and leaders of the most successful corporate partnerships, and collecting more data than I can reasonably summarize in any single keynote, I have arrived at a conclusion that surprises many people when I say it out loud:

Technology is rarely the differentiator.

The companies that win—that build Partner of the Year recognition, scale through marketplace, earn the trust of hyperscaler field teams, and grow when others are shrinking—win because of how they partner. They have the right mindset. They make a real commitment. They maintain their focus. They tell a helpful story. And they relentlessly pursue results, demonstrating an ability to adapt when the rules change.

In short, they win because of the principles highlighted in these pages.

I have known Vince Menzione for years, as a fellow voice in the partner ecosystem, and as a podcast and live event host who has brought me into more conversations than I can count. For four decades, he has served as one of the most credible and consistent guides for people navigating the corporate partnership space.

When I say credible, I mean it. Vince built a channel from scratch at a company on the brink of bankruptcy and turned it into the growth engine that sold the business. He led a $4.6 billion partner ecosystem at Microsoft. Then he went back inside as a channel chief and watched, from the other side of the table, how organizations struggle with the very problems he had spent his career solving. He built Ultimate Partner to help close that gap.

Vince's combination of skills—practitioner, executive, advisor, and community builder—is rare. And it is what makes this book different from the many others about partnerships that have been written by people who have studied it from a distance.

This book was written by someone who has been in every kind of room. And Vince has spent this past decade, through his podcast and live events, making sure that the most important conversations in our industry are accessible to every leader who needs them, and not just the ones lucky enough to have a seat at the table already.

The data I track suggests we are at an inflection point unlike any in the history of our industry. Marketplaces are the new channel infrastructure. Ecosystems are the primary growth engine. AI is rewriting the rules of every partnership, go-to-market motion, and customer conversation across industries and the world. And the organizations that have already aligned to these realities are pulling away from those that have not.

The Seven Principles laid out in these pages are an operating system designed to help you navigate this new reality. They are not tactics that will prove obsolete in 18 months. They are enduring. Vince's principles will guide you through this decade and beyond because they are grounded in human behavior, organizational dynamics, and the timeless economics of good partnerships.

Approach these ideas as a practitioner by putting them into practice. And when you find yourself in a room with the right people, doing the right work, building the kind of partnership that actually compounds over time, you will know exactly where your success came from.

Being in the right room changes everything. Vince has spent his career building those rooms. This book is his invitation to you to enter.

-Jay McBain Chief Analyst–Channels, Partnerships & Ecosystems Omdia

INTRODUCTION

SUCCESS
IS NEVER BUILT ALONE

Monday mornings, my team and I go through a vigorous exercise called "Ninety." It's an entrepreneurial process we use to ensure we're all on the same page as we kick off the week.

With a busy morning ahead, I knew I didn't have time to get to the gym, so I opted for a bike ride instead—a quicker way to get my cardio in before the day took off. I decided on my usual morning loop, out to the beach and back along US A1A, the coastal highway in Jupiter, Florida. Riding always cleared my head and got me pumped up to start the week.

As I pedaled, I thought through the key points of our weekly team call. We were only weeks away from our biggest event so far: the Ultimate Partner LIVE Summit in Redmond, WA. And this year, we were uniquely hosting it on the Microsoft campus, at the same conference center where I had hosted partners many times during my years as an executive at the hyperscaler. As a result, this event felt like a homecoming.

We were anticipating a record number of attendees, as the line up for this UP LIVE event promised Microsoft's Channel Chief along with several other top hyperscaler executives and industry experts. Our conference format is designed to weave in the Seven Principles, built upon a foundation of trust, that underlie any successful strategic alliance between two organizations aimed at mutual growth—and which provide the backbone of this book. The Ultimate Partner team had worked toward this event with singular focus and dedication for months.

I was waiting at the intersection near the end of my ride, the light about to turn green. As I pushed off, I felt ready to tackle the day.

That's the last moment of the ride I remember.

My next memory was of waking up in an ambulance on the way to the hospital. My adrenaline was racing. The EMT took my cellphone and called my wife to alert her, holding the phone up to my ear and mouth so I could let her know I'd be alright.

In the hospital, family friends greeted me, as did my wife. After a full scan, the doctors explained that I had broken three ribs, the orbital bone above my eyebrow, and my left ankle in two places. I was lucky not to be paralyzed.

How could this have happened?

I learned that an 88-year-old man making a left turn had driven his car straight into me. He told the police that he hadn't seen me. The impact knocked me to the ground unconscious, and the strap of my helmet pulled tight, cinching off my airway.

The police report also told of my extraordinary stroke of luck, as a woman driving by witnessed the accident. She happened to have EMT training, so she pulled over and sprang into action. She freed my neck from the helmet strap, cleared my airway, and got me breathing before the ambulance arrived. That woman saved my life.

I was lucky with my injuries. I was not so lucky with the timing.

While at the hospital, my wife made a call to our daughter Allie, who was getting married in just two months, to tell her what had happened. I couldn't even get up to go to the bathroom, let alone walk anywhere else. But from the bed, I yelled loudly enough for my daughter to hear me: "Don't worry, Allie. I'll be okay to walk you down the aisle!" I meant every word of it.

Meanwhile, in a matter of weeks I was supposed to stand on stage in Redmond leading a multi-day event filled with sessions led mostly by me. The UP LIVE event could not go on without me, and my team was concerned that I wouldn't be able to make it. But there was no way in hell I wasn't going to lead the event.

I had a steadfast determination to honor both commitments: to my team, our partners, our sponsors and speakers; and to my daughter. I knew I would keep them all. I didn't even think about the alternative. My whole life, I've been dedicated to the belief that I can achieve any goal, no matter how hard.

The recovery period wasn't easy, however. Between the trauma to my brain and the pain of my broken bones, I struggled just to stay awake. During the initial stretch, I couldn't get out of bed without help. My wife, Jackie, bathed me, fed me, and turned me over. Without her, I don't know how I'd have made it through. I started calling her "Nurse Jackie", and the name stuck. I was deeply grateful to have her by my side, and grateful that I had survived.

I took meetings with my team and calls with our speakers and sponsors from my bed — often with the camera off, too unsightly to be seen. My team did an amazing job filling in while I struggled, rallying to run the company and prepare for the event in my absence. I trusted them, and they delivered.

As news of my accident spread, support from my colleagues and community poured in. Along with visits from family members and friends, I received messages and calls from leaders whose stories I had shared on the Ultimate Partner podcast, people who'd flown in from across the globe for our UP LIVE events, and colleagues I'd worked with in the past at Microsoft and elsewhere.

Their support meant the world to me. It fortified my determination to walk across that stage in Redmond and then walk my daughter down the aisle. Slowly, with every physical therapy exercise and that visualization front and center, I began to take steps by myself again.

The date of the event arrived in a flash. Every minute of the nearly seven-hour flight from South Florida to Seattle was agonizing. I had to keep my ankle propped up, packed in dry ice packs. But I made it there.

That week in Redmond, sleep came in fragments. Every few hours, the pain and swelling in my leg would wake me, and I'd call down to the hotel bellman for more ice. At some point in the midst of one of those aching middle-of-the-night moments, icing my swollen ankle, it hit me: Ultimate Partner had stopped being just a business to me. I understood, for the first time with total clarity, that I was here on earth for a purpose bigger than any single event or company. My calling was to show up for this group of people, my community.

I ended up running the event, my ankle in a boot, walking back and forth to the stage with one crutch under my arm and my son John standing by my side. Backstage, he helped me move around, staying close and barely saying a word. I could feel his genuine care and concern. Thanks to him, I survived the event and days of sessions. His devotion deeply touched me.

A month later, it was time for Allie's wedding. By now, the boot was off and I had been going through extensive physical therapy to be able to walk her down the aisle. Since the time she was a baby, I held a vision of walking my beautiful daughter, my brown-eyed girl, to the altar. That moment when I was able to throw away the crutch and escort her down the aisle was one of the most important days and moving experiences of my life.

When I'd finally made it through that challenging experience of recovering from my injury and had time to reflect, it dawned on me that it was the most visceral, personal proof possible of a principle I've been teaching for nearly a decade: success is never built alone.

Not only that, but the family, company, and community support that had carried me through pointed to a deeper purpose in my career and life. Underneath the corporate partnerships and technology, I've always been driven by helping people achieve more in their lives. I'd worked with hundreds of leaders over the past decades, executing four wildly successful business transformations, and I knew there were many more individuals who could benefit from learning how to achieve their greatest results through the principles that we shared and the conferences that we led.

That moment is what led me to this book and to you.

Navigating a Shifting World

The incredible shifts happening in our world today lend a real sense of urgency to my message, particularly the rapid adoption of AI. Over the 40 years I've spent in the technology industry, I've never seen it change as fast as it's changing now.

Jay McBain, the analyst with Omdia who wrote the foreword to this book, spent his career spotting trends before the rest of us even feel them. His read on artificial intelligence is blunt: he calls it this generation's extinction event. Half of today's Fortune 500 companies, he says, will be off the list within 20 years.

You can feel it, and maybe that's why you picked up this book.

Many of the 17 million people in our partner ecosystem feel paralyzed at this moment. Perhaps you've lost your job or it has evolved into something new, or maybe you're realizing that you have an opportunity to expand your role in the midst of the shifts taking place. Some of you haven't fully absorbed the consequences of AI yet, but you're about to find out that the strategies that worked for you last year could evaporate this year. With this book, I aim to provide you with the kind of guidance that is crucial to navigate such immediate challenges.

What I realized in retrospect is that everyone who rallied to support me after my accident was already in place before the crash. My wife, Jackie, nursed me through my recovery; my team ran the company in my absence; my friends, family and community delivered messages of warmth and love. You can't build a rescue while you're lying in the road, struggling to breathe.

Companies are no different. The ones that survive shifts like those we are facing now commit to, build, and test their partnership strategies long before the day they need them.

I learned this the hard way. Years ago, a billion-dollar software company hired me, first as a consultant then later as an executive, to fix their partnership strategy. After I arrived, I discovered that I was their seventh corporate partnership leader in seven years. Soon enough, I understood why. The CEO continually stood in front of the investment community and pledged his support for partnerships. But behind the boardroom door, he funded none of it.

I've seen this more than once. Partner strategy cannot survive this kind of behavior. The real test of executive commitment lies in funding partnerships, measuring their impact, and rewarding them when they succeed. By that test, we had already failed.

But all was not lost. From that failure, I built the Seven Principles that sit at the heart of this book. The first three relate to taking action within your own company, your internal victory. The next three are about making an impact as you lock arms in partnership and in the marketplace. The seventh principle keeps you thriving even as everything around you changes.

The Seven Principles

Figure 1. The Seven Principles of Successful Partnering. A circular model showing how internal alignment drives external results, anchored by mutual trust and agility

Here is a brief overview of the principles we'll explore deeply in this book:

1. **Growth Mindset.** In a world of constant change, curiosity, learning and persistence are superpowers. Winners embrace possibility.
2. **Executive Commitment.** Success requires leadership buy-in and organizational alignment. It doesn't happen at the edges. It happens at the top—and cascades down.
3. **Vision and Value Alignment.** Great partnerships are built when leaders align on purpose, outcomes, and value creation. Without a shared north star, even the best-resourced partnerships will drift.
4. **Maniacal Focus.** The best results come when we prioritize, simplify, and concentrate energy on what matters most. Most partnerships die not from bad strategy but from scattered execution.
5. **Brand and Story.** Your story is your partner magnet. Partners and customers rally around belief, credibility, and trust. It is the fastest way to earn mindshare with your partner's field teams.
6. **Delivered Outcomes.** Outcomes are the currency of trust. Winning partners obsess over execution and measurable impact. Everything else is preparation.

7. **Agility.** Leaders must anticipate, pivot, and leverage change before it overwhelms them. What got you here will not get you there.

Trust stands at the core of everything we discuss, and determines how much each principle pays off. That's why I call it the multiplier. You can only build it over time, with dedication and effort. Yet when you do, it rewards you when you need it the most, just as the people I trusted showed up for me in the hospital in the days following my accident.

The Path to Powerful Partnerships

If partnerships—whether with Microsoft, Google Cloud, Amazon Web Services (AWS), OpenAI, Anthropic, or another platform—are central to your company's success, then this book is for you. It will show you where you stand in terms of your alliances, what to fix first, and how to grow from millions to hundreds of millions of dollars in size as a result.

Every chapter draws from my decades of experience across startups, a turnaround, nine years at Microsoft, and the work we do every day at Ultimate Partner. It also draws from the collective wisdom of a community that has shown up over the years to learn, challenge assumptions, and share what actually works for them.

Throughout every chapter, you will hear not just from me, but also voices from the over 300 conversations I have had with the most respected leaders in the partner ecosystem through the Ultimate Partner podcast. I want to emphasize the extent to which I consider these individuals the co-authors of my Seven Principles. They include: Jay McBain of Omdia, who wrote the Preface; Tony Safoian, CEO of SADA, one of the most successful partner organizations; Dr. Michael Gervais, one of the world's leading performance psychologists; Dai Vu, General Manager of Google Cloud Marketplace; Janet Schijns, CEO of JS Group; Cyril Belikoff, Microsoft's VP of Marketplace; Matt Yanchyshyn, VP of AWS Marketplace; Dux Raymond Sy, Chief Transformation Officer of AvePoint; and many other leaders in the field.

One other thing I want to make clear before we dive in: this book is not about tactics for building powerful corporate partnerships. Tactics shift constantly. The Seven Principles, on the other hand, remain steady no matter how much industry turbulence surrounds them.

Again and again, I've watched capable, hardworking leaders lose their way in the partnership ecosystem. Their problem was never effort or implementation of the latest trendy strategy. It was orientation around principles. They moved fast, with no fixed point to navigate by.

There's a historic lighthouse at Jupiter Inlet near our house. I biked past it that fateful morning of my accident. Since 1865, it has weathered every storm, helping sailors navigate the waters of one of the most treacherous inlets along the Florida coast. It does so by holding steady, keeping its light burning bright, and trusting that the people who need it most will find it.

I'm grateful to be alive today. My brush with death inspired me to write this book. I developed the Seven Principles to serve as a beacon to those of you who are already in the Ultimate Partner community, as well as those of you who are ready to join us.

As the market continuously spins, when you can't tell up from down, may this book serve as the guiding light you can depend upon to inspire you to keep doing your best work and deliver you to safety..

CHAPTER 1

YOUR STRATEGY MAY ALREADY BE OBSOLETE

During my 40 years in this industry, I've lived through four major disruptions: PC, internet, mobile, and cloud.

Each one felt like an earthquake while it was happening, changing the very nature of what we offered and how we worked with our partners. Yet each shift eventually settled. We held on and adapted. Then we got back to business, finding our footing on ground that had changed shape.

Somewhere in your organization, there's a plan everyone worked hard on, everyone approved, and everyone believes in. It probably has AI all over it—a roadmap, phases, a target state. Underneath it lies a premise no one has spoken out loud:

that this earthquake will pass, the way the others did. The market will settle. There's an "after" coming, so you just have to hold on until it arrives.

The problem is this time, there is no "after."

The Six Shifts

The earthquakes rocking our industry and the world right now don't have an ending. We're being hit by six massive shifts at once: an increase in the sheer pace of change, geopolitical uncertainty that's eroding trust, an economy that's demanding we do more with less, buyers who don't buy the way they used to, a handful of platforms emerging with more gravity than the rest of the industry combined, and AI that's rewriting the rules about what customers will pay for and why.

Each shift impacts the others. And no company, no matter how big or well-funded, will get through them alone. The ones that win will do so in partnership.

Let me take you through these six shifts and what they mean for you.

Shift 1: The Accelerated Pace of Change

The first shift is not about technology itself so much as the pace of change.

In March 2020, when the Covid pandemic hit, the world stopped. Tragically, we lost my mother before it was over.

Everyone on earth was forced to adapt to massive change faster than any of us had experienced. How we work, shop, educate, and deliver healthcare transformed almost overnight.

As Microsoft CEO Satya Nadella put it, "We saw two years of transformation in two months." Ever since, that velocity of change has become the new constant.

The implications for partnership strategy are profound. A plan created for a 12-month horizon may be obsolete in six. Winning organizations have to build adaptability into their operating models.

Shift 2: Geopolitical Uncertainty and Eroding Trust

Trust is the foundation of any successful partnership. This is why I reference it repeatedly. But right now, trust is under pressure everywhere. And this is the second shift we're facing.

We have moved from the "flat world" era of globalization to one characterized by geopolitical risk, regional conflict, political polarization, and a deep erosion of trust in institutions from governments, to media, corporations, and each other.

The threats for technology infrastructure companies have never been more complex, with nation-state level bad actors and criminal organizations targeting the systems we rely on to run our businesses.

For partnership leaders, this matters in two ways. First, it creates new complexity in the supply chains, regulatory environments, and compliance frameworks that underpin enterprise technology decisions. Second, and more importantly, it takes trust within your ecosystem from a nice-to-have to a genuine competitive differentiator. The partners who have built genuine credibility over time are the ones buyers will turn to when uncertainty is high.

Shift 3: The "Do More With Less" Era

The Covid technology boom was followed by a sharp pullback, in part driven by AI. The layoffs that began in 2022 to 2023 at Google, Amazon, Microsoft, and other major tech companies have become the new norm. Hyperscalers are reducing headcount, and partner-facing roles have been cut disproportionately.

The philosophy of "do more with less" appears to be the new permanent operating posture. For many partners accustomed to robust vendor support, the message is clear: drive your own results. Invest in your own capabilities, build your own community relationships, and develop the internal expertise you need to navigate without relying on the hyperscaler to do it for you.

Shift 4: Radical Changes in Buying Behavior

Rodney Clark, a former colleague of mine at Microsoft who went on to run partnerships at Cisco, once told me, "It is the customer buying behavior, the shift in customer expectation, that has created the need for all of us to rethink our models." Not technology, hyperscalers, or AI, but the customer.

Something happened while most vendors were not paying attention: buyers got smarter than sellers. This is the fourth shift.

Today's enterprise technology buyers do their own research, consult their own trusted advisors, and form their own points of view long before they engage with vendor sales teams. The cold call is dead. So is the unsolicited email. The product demo as an opening move is dead, too.

And the new demographic reality accelerates all of it. By 2030, an estimated 75% of technology buyers will be Millennials and Gen Z. These buyers have fundamentally different expectations. They prioritize verified outcomes over brand names. They trust their inner circle over vendor marketing. And increasingly, they would rather complete a transaction through a cloud marketplace than sit through a sales cycle.

That last point matters more than most organizations realize. Cloud marketplace revenue is on track to represent 32% of all B2B software revenue this year. The buyer who prefers to transact through a marketplace used to be an outlier. Now they are the future of enterprise procurement. The committed cloud budget sitting inside Microsoft, AWS, and Google Cloud agreements is not waiting for your sales team to find it, but rather looking for your marketplace listing to surface.

The buyer has changed, and the channel that reaches them has changed with it.

Shift 5: The Rise to Dominance of Platforms

Today, Microsoft, AWS, and Google Cloud each operate hundreds of massive data centers globally. Their combined investment in AI infrastructure is well into the hundreds of billions of dollars annually. Joining them are Anthropic, Nvidia, OpenAI and other potential players in the artificial intelligence landscape.

The capital required to compete in this space is so enormous that only a handful of organizations can play. What this creates, for everyone else in the ecosystem, is a gravitational reality: a handful of organizations now anchor the go-to-market strategy of virtually every enterprise technology company in the world.

You do not partner with hyperscalers because it is a good idea, but because you have no choice. This is the fifth shift.

Kevin Peesker, former President of Microsoft's Worldwide SMB & Corporate Markets—a $60 billion global business—has said, "The CEO is the new CIO." Technology purchasing decisions have migrated from IT departments to boardrooms. CEOs are signing multi-year commitments, and hyperscalers have aligned their entire go-to-market around that reality.

For partners, this changes everything about how you build relationships with, sell, and measure success, and what your value proposition needs to say.

Shift 6: The Emergence of AI

I promise you this is not an AI book.

But if I wrote about how to succeed in the partner ecosystem given the massive shifts impacting our industry without giving AI its full weight, I would fail you. AI runs through every chapter of this book because it is vital to every aspect of business today.

It's just that AI itself is not the point. The Seven Principles, which serve as the compass helping to guide you through these shifts, are.

As Dux Raymond Sy, Chief Transformation Officer of AvePoint, put it, "What happened with OpenAI is that it just democratized AI. ChatGPT made AI available to the public, just like the World Wide Web did for the internet."

AI is being built into everything we do. It is, undeniably, shaping our future. And it's moving at warp speed. "This is the era of the fastest technology transformation in history," commented Nicole Dezen, Microsoft's Chief Partner Officer. "What took cloud 15 to 20 years will happen in five."

I was in rooms with partners, leaders, and executives as these events were unfolding. The conversations changed overnight. Every meeting and conference became, in some way, a conversation about AI. From that moment forward, nothing in the partner ecosystem has been the same.

Three Companies, Three Choices

Satya Nadella recognized this inflection point immediately. Microsoft had already been investing in OpenAI, and when ChatGPT launched, the company moved with a speed and conviction I've rarely seen from any organization, much less one at scale. They poured billions into the technology embedded into products, and quickly launched Copilot as the AI interface for the entire Microsoft stack.

By 2026, the race had opened up. Google came to play, replacing its Vertex AI roadmap with the more ambitious

Gemini Enterprise Agent Platform. It was built to deploy and govern production agents across Google Cloud, other clouds, and on-premises systems. Now Google is building for a multi-cloud world, with its frontier Gemini models as the draw.

AWS made a different bet entirely. Rather than leading with its own models, it built Bedrock as a model-agnostic platform. In April 2026, it expanded its partnership with OpenAI to bring in their models and managed agents to Bedrock. AWS positioned itself as the operational foundation for enterprise AI, focused on getting you from prototype to production regardless of which model you choose. The bet is working. AWS reports Bedrock processed more tokens in the first quarter of 2026 than in all previous years combined.

The three hyperscalers made three strategically different plays. Microsoft bet on deep product integration, Google on its own capabilities and cross-cloud orchestration, and AWS on infrastructure scale and model optionality. Underneath all three, the model companies—OpenAI, Anthropic, DeepMind, Meta AI, and others—are well-funded, seriously staffed, and running hard in the same direction.

It is impossible to predict which model will win. Frankly, it is not even clear whether there will be one winner.

Here is what matters more than who wins the horse race, however. We are no longer in the AI experimentation phase. We are in the accountability phase. As Nina Harding, who leads Microsoft's partner business in the Americas, made clear, customers are now asking one question: "What will your AI solution deliver for a company like mine, in my industry, with my specific business problem?"

The partners who answer this question with hard evidence and industry depth will succeed. The ones still running pilots without ROI stories will not.

And increasingly, the evidence is being interpreted by AI as well as people. One partner CEO in our community put it this way, "The machines need to verify via trusted third-party data sources what it is that you're up to. How many deals have you done? What is your win rate? This kind of information is incredibly valuable to the agent world."

The prize for getting it right is enormous. Canalys projects $158.6 billion in partner revenue tied to AI by 2028.

For partner organizations watching this race, early commitment to a single platform or model is genuinely perilous. Those that locked in before the agentic layer began reshaping everything are already navigating the negative consequences. The landscape shifted faster than most of us predicted, and it will shift again.

So, my advice is to remain "solution agnostic" wherever possible. Build your AI capability on the Seven Principles, which do not shift: growth mindset, executive commitment, alignment on vision and values, maniacal focus, a powerful brand and story, delivered outcomes, and agility, all held together by trust.

The partners who win will bet on principles, not horses.

Cyril Belikoff, Microsoft's VP of Marketplace, provided an analogy that beautifully captures the present opportunity for partner organizations. "Thomas Edison invented the lightbulb. It changed life in the home and in the workplace. Then people needed sources for that lightbulb—infrastructure, power, power stations. And then people started to innovate. Not just lightbulbs, but dishwashers and washing machines.

"AI is at that same platform level today," Cyril continued. "We are building a few lightbulbs. But we need partners to build the dishwashers and washing machines—to innovate across the new opportunities for customers."

This is the opportunity of the AI era. The aim isn't to compete with Microsoft, Google, or AWS by replicating what they build. The aim is to develop the applications, integrations, industry solutions, and managed services that make AI useful in the real world—for real customers, with real problems, regardless of which platform they run on.

The Shifts Themselves are Shifting

What makes this moment different from every previous era of disruption is not only the magnitude of the changes just discussed, but their very nature.

They are happening simultaneously. AI, platform dominance, new buying behavior, geopolitical pressure, and the rest are arriving not one at a time, in a sequence, but all at once. To the same organizations. Requiring the same leaders to respond at the same moment.

They are interdependent. You cannot align to the marketplace priorities of Microsoft, AWS or Google Cloud without understanding what AI means to their field motions. You cannot build a co-sell strategy without understanding how the buyer has changed. Every thread connects to every other thread. Pull one and the others move.

They are ongoing. I told you at the start of this chapter that there is no "after." Now you have seen why. There is no "new normal" on the other side of these earthquakes because they are ongoing. The disruptions will never stop coming. The organizations that treat these shifts as temporary and try to wait them out will therefore fall further behind with every quarter. You must respond quickly, before your foundation crumbles.

This is why I developed the Seven Principles in the first place. They serve as your compass, designed to guide your path forward. When the market moves, so do your tactics. But your principles hold steady, so that you can return to them whenever you need, even in the midst of constant change.

It's time for you to implement these Seven Principles so that your business can continuously adapt and increasingly prosper.

YOUR TURN

Look at the shifts described in this chapter—the rapid acceleration of change, lack of trust, "do more with less" mentality, new buyer behavior, platform dominance, and the AI accountability era.

Answer these questions honestly:

1. *Where does your organization stand? Are you building for today's world? Or are you still building for the one that existed three years ago?*
2. *Where are you getting your most trusted guidance on navigating these six shifts? Is it from the hyperscaler's partner portal, from independent communities and voices, or from peer relationships you have built over time?*
3. *If your current partnership strategy is built for 2021, what is the single biggest thing you can change to make it better for right now?*

YOUR COMMITMENT THIS WEEK:

Identify one assumption in your current partnership strategy that has not been revisited in the last 12 months. Schedule time to pressure-test it against the current market reality.

.

CHAPTER

2

WHY PARTNERSHIPS FAIL

The courtship lasted months.

By 2018, I had left Microsoft, started my podcast, and begun consulting to future top partner companies. One of my clients was a large, established technology company—and they wanted more than my advice. I'll call it Salience, though that's not the organization's real name. Eventually, they flew me in to meet the executive team and put an offer on the table: Vice President of Channel and Ecosystem.

Saying yes meant walking away from clients I had just started to build, uprooting a comfortable life split between New Jersey and Florida, and moving my family across the country. But the prize was real: impact at scale, a seat in the boardroom, a path to one of the most senior roles in the company.

Saying yes also meant breaking my own rule. As a consultant, you are seen as the smartest person in the room. The day you become an employee and start wearing a corporate name badge, the same people start treating you differently. It's as though suddenly you're not so smart anymore.

Nevertheless, I believed in the opportunity. I had done my due diligence. So I said yes.

And for a while, the results said I was right. I drove Salience into the top tier of the Microsoft ecosystem. We won Partner of the Year as its number one co-selling partner. I put the company on Microsoft's radar in a way it had never been before.

That's the shiny surface of this tale. Dive in, and you'll see the muddy mess underneath.

Every month, I walked into the executive team meeting and laid out a plan for building more revenue through partnerships. Every month, they replied with the same enthusiasm. And every month, the resources failed to arrive.

I found one ally, the Chief Operating Officer. The two of us spent many long nights at the whiteboard, yet we made little headway. The Chief Revenue Officer stood forever in our way. You can feel it when a person is actively working against you. He'd contradict our ideas, ask questions designed to engineer stall-outs, and above all, deny us the funds we needed. He was a wall we could never climb over.

Above us all sat a CEO with a friendly board behind him. Time and again, he said, "I believe in partnerships. We're building a partner strategy. We brought in a leader from Microsoft, and he's going to get us there." Yet he didn't back up his words with action. Behind the boardroom door, Marketing had no budget, Product had no roadmap, Sales had no directive and professional services delivered directly. Not one function carried Partnerships in their plan.

The numbers told the real story. Salience fielded hundreds of direct salespeople and multiple marketing organizations, including dedicated teams for events and digital. These were funded at a scale that made the executive commitment to the direct-selling model unmistakable. Meanwhile, my Partner team consisted of ten people, operating on a fraction of the marketing budget alone.

Then one day, someone mentioned offhandedly, the way you'd mention a change in the weather, "You're the seventh partner leader to hold this job in seven or eight years."

Yikes.

Later, my predecessor called—half to welcome me, half to warn me. But by then, my wife and I had already moved across the country and signed for the house.

I did everything a person can do to not make the same mistakes. I created cross-functional teams and pulled leaders from every department and vertical into my weekly and monthly meetings. I planted the partner strategy throughout the organization by hand. Still, it would not take root. You cannot personally compensate for a company that has cut off your oxygen supply.

After two years, Covid hit and offered me the perfect exit. My mother died months later. The job had taken away my chance to spend more time with her, and the virus took away my chance to give her a proper goodbye.

For a long time, I wore that experience like a scar. I don't anymore. It turned out to be the most valuable failure of my career, and it is a big reason why this book exists. The people at Salience were not villains. They were simply part of an organization that had never actually made the decision to support partnership.

Everything I have built since—the podcast, UP LIVE events, Ultimate Partner community, and this framework—began with the need to understand why a company that said all the right things could not execute on any of them.

Above all, I share this unfortunate saga because throughout my years of working in the partnership arena, I've learned that it is not unusual. It is, in fact, the pattern.

I hear some variation of the same story all the time. A company hires a sales or partner leader to change their strategy. But just as the new leader begins building a terrific program and team, the senior executives pull the plug, hit pause, or scrap the program entirely.

This happens not because of a market failure or a technology failure, or even because something was off with the hyperscaler program. It happens because something wasn't functioning well internally—something embedded so deeply within the organization that no strategy or program could reach it.

The Four Failures

In my experience, four deeply embedded failures most commonly block success before Partner strategy ever gets a chance to work. These are: a toxic culture, lack of commitment, poor communication, and mistrust.

These four failures lie at the heart of what goes wrong when organizations say they believe in partnership and then behave as if they do not.

I lived all four at once while at Salience. I've also witnessed them from the hyperscaler side while at Microsoft, on the partner side, and on the advisory side at Ultimate Partner. Let me show you each, so that you can recognize them in your own organization before they cost you another partner leader.

Janet Schijns, CEO of JS Group and a recurring voice in the Ultimate Partner community, put a sharp point on the long-term consequences of these failures: "The companies that will fall off the Fortune 500 list in less than 20 years are the ones who don't elevate partnering to the C-suite." She was talking very specifically about whether organizations structurally understand that most of their revenue already flows through partners, and builds accordingly.

Those are the stakes for partnering. Now here are the problems that frequently get in the way.

Failure 1: Toxic Culture

Culture is the soil in which partnerships grow. It can be either fertile or toxic.

When Satya Nadella became CEO of Microsoft in 2014, he recognized immediately that the culture had turned toxic. Why? Because internal teams were competing against each other for resources, recognition, and credit. Stack ranking rewarded individuals for outperforming their colleagues rather than building together. The result was a company where internal competition destroyed external opportunity.

Nadella determined that it was time for a massive cultural transformation. He spearheaded a new mindset—one that encouraged collaboration and rewarded teamwork. Then he reinforced this culture with every structural decision the company made. Transformation happened because leadership changed what it valued and prioritized. That standard applies as directly to your partnership culture as it did to Microsoft's.

Heroic Sales Culture

The most corrosive cultural force I have encountered in partnership organizations is a heroic sales culture. It is born when an organization builds its identity around individual sellers—the people who close the big deal alone, ring the bell, and take the trip.

It is toxic because heroic sellers do not want partners in their deals. Partners mean shared credit. Shared credit means a smaller commission, no trophy, and no story to tell. So they tell customers the solution is proprietary. They register deals to block channel involvement.

They do what the culture rewards them to do. And the culture protects them.

In fact, in a lot of organizations, problems with corporate culture evolve directly from heroic salespeople. We give them so much credit that we tend to leave them alone. It's less tolerated these days than it used to be, but I've seen CEOs protect top sellers when they got drunk at work, did something stupid in front of everyone at the annual conference, and even engaged in sexual harassment.

Over time, the heroic seller culture becomes the sacred cow, embedded so deeply in the organization's DNA that no partner program, executive mandate, or training initiative can uproot it. And it's how partnerships die.

Stephen Covey called the underlying belief of this sort of organizational culture "the scarcity mentality." It forms on the conviction that there is only so much success to go around. "When I win, you lose. When I share, I lose."

This attitude is the opposite of what successful partnering requires.

The antidote is building an abundance mentality. From the leader on down to every sales person on the team, the organization must promote the belief that partnerships pay off. One plus one equals three or more. Leaders model the belief from the C-suite, reinforce it with compensation decisions, reflect it in performance reviews, and embed it in the stories they tell about what success actually looks like.

Failure 2: Lack of Commitment

Organizations that fail at partnership often also lack executive commitment. And as with a toxic corporate culture, you can usually tell within weeks that this is the case.

These executives invest in a partner team. They take meetings with hyperscalers. They even publish a partner program and announce it with fanfare. They talk the talk at partner conferences and in boardroom presentations.

But when it comes to taking the steps that actually create results—modifying compensation to incentivize co-sell, allocating real budget to partner marketing, giving the Channel Chief genuine access to the executive team, and giving the program enough time to gain traction—they fall short.

I learned this lesson the hard way, and early on in my career.

Before 9/11, I was asked to build the federal government channel for a small company called Itronix that made rugged notebook computers. These were purpose-built machines designed to survive field conditions—the kind of environments where a standard laptop would not survive. We were a startup in turnaround mode, and I had been brought in to build a business that did not yet exist.

The CEO who hired me thought six months was enough time to see results. He did not understand how the federal government actually works. Before a single dollar moves, you need a GSA schedule, which is the prenegotiated contract that lets agencies buy from you. This takes time. You also need access to buyers and influencers, and that takes time. On top of these constraints, the government does not run on your fiscal quarters. It runs on the Congressional calendar.

While the CEO wanted revenue right away, I needed a runway.

Thankfully, the private equity firm behind Itronix understood this, so they replaced him. The new CEO got what I was building and made a genuine commitment of time and resources. He believed that the strategy would compound if given room to breathe.

I took that commitment and ran with it. I got rid of the people who were not serious and became deeply involved in selling myself. I mapped every relationship that could open a door, and pursued each one with everything I had.

As an example, we wanted to sell to the U.S. Army. I found out that a multi-billion-dollar defense contractor held CHS-3, the field-support contract for Iraq and Afghanistan, forward-deploying people in theater to swap out failed computers. If you wanted to sell to the forward-deployed Army, you had to buy from them.

Just one problem: they were already working exclusively with our competitor.

Digging around, I discovered the key decision maker was a fellow named Chris Marzilli. But I couldn't get Chris to take my calls. Actually, back in those days, you didn't even stand a chance of getting a senior executive on the line if you were an unknown. You had to pass through their assistant first.

So I took a big risk. I called Chris at 7 AM. Just as I'd hoped, he answered his own phone. "You need to meet with me," I said. He listened.

Then I went to work. I connected Chris with the congressman in Spokane, Washington, where Itronix was based. We flew with Chris to Washington, DC and put him in front of the people who controlled the funding he needed. Our local congressman got behind it. The funding came through.

My abundance mindset is how our partnership was made real.

Then Golden Gate Capital put us up for sale, and two companies bid for us. Post-9/11, with the U.S. fighting two wars, a defense giant with its own Gulfstream fleet was competing to acquire a small rugged computer company from Spokane that no one had heard of a few years earlier.

The day it closed, the General Dynamics executive who led the deal said something I have never forgotten. "We bought Vince."

He was not paying me a compliment. He was describing an asset.

None of this would have happened without the second Itronix CEO making a full commitment to partnerships. Not the Army relationship. Not the General Dynamics acquisition. Not the compounding results that turned a struggling startup into a company worth fighting over.

When organizations suffer from this dysfunction, they end up cycling through sales leaders and channel chiefs every 18 months, each time declaring the previous strategy a failure.

They blame the problem on talent when the real issue is commitment.

True commitment can show up in several ways: allocating the right resources and real capital, aligning the organization in such a way that gives the partner leader genuine authority, allowing enough time for the partner strategy to take hold, and demonstrating a genuine willingness to change the business model.

Failure 3: Poor Communication

The third failure is poor communication.

Most organizations only show up when they need something. The best ones show up consistently, share their wins, and treat their hyperscaler counterparts like partners rather than transactions.

Every partner leader knows the pattern. Two organizations come together. The first meeting is electric with shared excitement, aligned language, and abundant enthusiasm. Everyone leaves energized.

And then… silence.

People from both organizations go back to their respective desks. Even with the best of intentions, behavior does not change. The internal rallying cry is not sustained. External communication becomes inconsistent. The follow-up that was supposed to happen in two weeks happens in six, or twelve, or not at all.

Many partnerships die in this gap between intention and reality. Closing it takes what I call being "diplomatically aggressive." Not waiting for the other side to do the work, but showing up, consistently and purposefully, thereby creating a feedback loop that keeps both organizations moving. It means being the one who sends the recap nobody asked for, and the one who calls in week two instead of week six.

The best partners I have observed treat communication with their hyperscaler and ecosystem partners the way a great account executive treats a top customer: with regularity, substance, and a genuine interest in understanding what the other side needs to succeed.

Failure 4: Mistrust

The final failure is lack of trust. Without trust, nothing else in this book works.

At Salience, the high degree of organizational mistrust became obvious when an employee mentioned to me in passing that they had burned through seven partner leaders in seven years.

Trust underpins all Seven Principles, which is why it gets its own chapter.

Awareness is Power

These four failures did not come from an ivory tower or a theory I invented. They came from watching what the organizations that broke through actually did differently or better. I've had four decades in this business, conducted more than 300 podcast conversations and more than a dozen Ultimate Partner LIVE events. I've worked with every partner type across every platform ecosystem. My analysis emerged from the evidence.

Notice the thread running through all four failures. Toxic cultures persist because nobody answers for what gets rewarded. Hollow commitments persist because nobody holds executives to them. Communication dies because nobody owns the follow-up. Trust erodes one unkept promise at a time. Wherever you find these failures, accountability is missing at the scene.

The good news about the four failures is that they are not a death sentence. On the contrary, they are identifiable and addressable. Every organization I have worked with has suffered from at least one of them. Most have suffered from more than one. The best companies found a way through.

Understanding the failures is the first step to eliminating them. Then, you can work to replace them by implementing the Seven Principles.

But the work begins with an honest assessment of where you are today.

YOUR TURN

The four failures are rarely visible from inside the organization. They more often feel like external obstacles. People are inclined to blame the market, the hyperscaler, and the competition when the real constraint is internal. This exercise is designed to make the invisible visible.

1. Rate your organization on a scale of one to ten in each of the four areas, with one being lowest and ten being highest. Where is the lowest score? That is your most evident failure.
    - Culture: ____
    - Commitment: ____
    - Communication: ____
    - Trust: ____
2. Think about the last major partnership initiative that stalled or failed to deliver. Which of the four failures was most present? What would you do differently today?
3. If a new channel chief joined your organization tomorrow, what would they discover in the first 90 days that you already know but have not fully addressed?

YOUR COMMITMENT THIS WEEK:

Have one honest conversation with a member of your leadership team about the failure you scored lowest on.

CHAPTER

3

THE MULTIPLIER:

TRUST

My grandfather came to Ellis Island with just 20 dollars in his pocket. His is a typical American immigrant story of the early twentieth century.

After World War I, my namesake left his small village outside Bari, in the Apulia region of southern Italy, looking for opportunity. He crossed the Atlantic Ocean in a steamship with little more than the address of a relative and a conviction that if you showed up, worked hard, and delivered on your word, this country would reward you.

And work hard he did. His first job was hauling giant blocks of ice up steep flights of stairs to people's "ice boxes" in an era before refrigerators. Backbreaking work, he never complained.

Without capital, connections, or a college education, Vincent built businesses that lasted. He did this by establishing a reputation—one day at a time, one transaction at a time, one kept promise at a time. People knew who he was. They understood what his handshake meant.

Trust was the foundation of everything my grandfather created.

I did not understand this because he talked about it. I simply absorbed the essence of who he was long before I even had the language to describe it: my grandfather was a man of his word.

My father, John Menzione, carried that same conviction into a different theater. He served in the U.S. Army in World War II. After he came home, he raised his family on this same foundation, which he lived every single day: Integrity is everything.

Two men of two different generations who operated based on the same core principle. They didn't need business school or a framework to prove it to them; they just knew. The currency that compounds most over time is trust.

I have spent four decades in business proving they were right.

Trust is Oxygen

Everything comes back to trust.

In nearly every podcast conversation I've had, when I ask leaders what they see in the best partnerships, they almost universally name trust as the key ingredient. And when I ask them to share the kryptonite that makes partnerships fail, it boils down to an absence of trust.

Trust is the oxygen of partnerships. Without it, everything else withers.

A shared vision means nothing without trust. A communication plan falls apart without trust. Executive commitment erodes without trust. Culture becomes toxic without trust.

And you can't fake trust, nor can you rush it.

My grandfather and father knew that integrity cannot be manufactured on demand. It cannot be declared in a press release or embedded in a partner agreement. Trust must be earned through consistency, transparency, following through on what you promise, and showing up even when it is inconvenient.

Dr. Michael Gervais is a world-renowned high-performance psychologist who has worked with Microsoft's leadership team, as well as the Seattle Seahawks and Olympic athletes. I had the incredible good fortune of getting to work with him while at Microsoft, and I've interviewed him on my podcast several times.

He says, "Change is not going to happen unless psychology happens first. The internal revolution has to precede the external results."

Trust is the foundation of that internal revolution. When an organization commits to building trust—genuinely, structurally, at every level—partnership results follow. When it does not, no strategy plan or investment will compensate.

Beyond the Seven Principles

The Seven Principles defined in this book act as an operating system. Each one builds on the others, and each one is necessary. Together, they form a complete architecture for partnership success.

I have put the Seven Principles in a specific order, and that order is the point.

The first three—growth mindset, executive commitment, and vision and value alignment — are internal. You must do the work inside your organization before you can successfully execute anything outside of it. I call these the Internal Victory.

The next three—maniacal focus, brand and story, and delivering outcomes—are external. They apply to how you execute in the market, build your reputation with hyperscaler field teams, and turn your efforts into compounding revenue. I call these the Partner Victory.

Agility wraps around the entire model. Trust compounds, revenue compounds, but so does irrelevance. The companies that stumble in this decade won't fail from bad strategy. They'll fail from good strategy executed a year too late.

Trust stands as the solid base, running through all of it. It is the connective tissue between the Internal Victory and the Partnership Victory. You cannot build genuine shared vision with a partner organization that you do not trust. You cannot execute with maniacal focus if your internal team does not trust the leadership. You cannot build a brand story that lands if the market does not trust that your outcomes are real.

I do not view trust as the eighth principle because it is vital to every single phase of the partner journey and every aspect of your organization. That is why I call trust the Multiplier.

Get it right and the Seven Principles compound. Get it wrong and even the best strategy, the biggest budget, and the most senior executive sponsor will not save you. Principles collapse. Your co-sell motion stalls. Your hyperscaler relationship stays surface-level no matter how many joint business plans you file. Your field sellers choose someone else's name when they have one call to make.

Without trust, you do not have partnership. In fact, you barely have a company.

You Cannot Manufacture Trust

When we launched the first Ultimate Partner LIVE event in November 2023, trust is what got us started.

Executives I had worked with at Microsoft did not have to return my calls. Sponsors did not have to support our event, especially before they had seen results. Community members were not obligated to fly across the country to Las Colinas, Texas to attend. Yet they did.

This did not happen because I had marketed the event well. It happened because over the course of decades, in hundreds of rooms, I had tried to show up for people the way I wanted them to show up for me. And so they trusted me.

I trusted them, too, so our conference became theirs. We built the room together.

Then, 18 months later, I was lying in that hospital bed after my bike accident, unable to move and uncertain whether I would make it to Redmond. And my community showed up again.

When a deal goes sideways, a hyperscaler changes its priorities, or a key executive leaves and the relationship has to start over, trust is either there or it is not.

You cannot build trust while you are lying in the road, battling for breath.

You cannot rush it. You cannot buy it. You cannot manufacture it in the moment you need it. You can only earn it, one room at a time, one kept promise at a time, one relationship at a time.

This is the part of partnerships that most organizations discover too late.

The Playbook Era

Over the last five to six years, a new class of voices has stepped onto the partnership scene: analysts, advisors, and content creators, many following the trail Jay McBain blazed. I am one of them.

Much of what we produce is a playbook. How to build a co-sell motion. How to use AI to do the work of five people. How to close the revenue gap. I am for all of it. The leaders in our rooms who have written their playbooks down are running deliberate motions instead of hoping something works, and that discipline shows up in their numbers.

But in order for a playbook to work, the willingness to partner must already be there in your organization. What I've witnessed is that the revenue gap is not a tooling gap. It's a trust gap. The reason hyperscaler's field sellers are not bringing you deals is because no one has convinced them that partnerships are worth the effort.

Usually that's because someone got burned. A partner overpromised. A co-sell motion produced meetings instead of revenue. The alliance looked good in the deck, yet delivered nothing.

No playbook will fix that problem. Neither will an AI agent. A playbook in the hands of an organization no one trusts only offers a faster way to fail.

The Trust Economy

We are moving from an attention economy to a trust economy.

In the attention economy, you won by being loud. Social media built an entire industry on volume and visibility, being everywhere all at once. Companies spent millions to get in front of as many eyeballs as many times as they could.

In the trust economy, you win by being believed. Your track record precedes you. You're the organization that hyperscaler field sellers, channel partners, and enterprise customers turn to when they need someone they can rely on. The person who randomly shows up in a LinkedIn feed doesn't get the deal. The person or company with a great reputation in the ecosystem does.

This shift is urgent right now because AI agents are increasingly the ones making recommendations. Moving beyond Google ads, they dive into Reddit feeds, earned media, and Trustpilot reviews. They examine your co-sell win rate, customer outcomes, and consistency of delivery.

It's getting a lot harder to game the system.

AI does not change what trust is. It changes what trust is worth. Companies that have built genuine trust over time and across a broad range of organizations are finding their voices amplified. Companies that have not are finding it increasingly difficult to be found at all.

My grandfather did not know what an AI agent was. But he understood, better than most, that the reputation you build when no one is watching is the one that determines everything.

YOUR TURN

Before you move on to the next chapter, I want you to sit with a question: If the hyperscaler field sellers who know your organization best were asked to evaluate, anonymously, whether they trust your team to deliver on what you promise, what would they say? Unfiltered and speaking honestly.

Your answer may be the most important data point in this entire book.

Building partnerships with the right people changes everything. But first, you have to earn their trust.

YOUR COMMITMENT THIS WEEK:

Find one promise your organization made to a partner that is still open. The introduction you said you would make, the follow-up that slipped, the answer someone is still waiting for. Close it before Friday.

CHAPTER

4

PRINCIPLE 1:
GROWTH MINDSET

"I'm really stoked," said Dr. Michael Gervais in his unmistakable surfer cadence, casually mentioning the upcoming session he had with CEO Satya Nadella and Microsoft's senior leadership team.

Michael is one of the world's leading high-performance psychologists, the person organizations call when they want to understand what separates world-class performers from everyone else. Already he'd worked with Red Bull, the Seattle Seahawks, Olympic medalists, and elite performers. Now he was working with us.

I was lucky enough to be in the room with the rest of Microsoft's Public Sector leadership team as we piloted Michael's high-performance mindset training for our company. What I experienced fundamentally shaped how I think about partnership success to this day.

At the time, none of us fully grasped what our team session with Michael would set in motion. Turns out, we were serving as the warm-up act for one of the most consequential cultural transformations in modern business history. I'm going to lean heavily on Microsoft in this chapter as a result. I do this not out of loyalty, but because I got to participate in this extraordinary event from the inside.

Years later, when I traveled to Los Angeles to sit with Michael at his Mastery Labs Studio for a podcast interview, I asked him what it was like to work with Satya's executive team in the months following our public sector team session.

"That leadership team was full of alpha competitors," Michael said. "Highly intelligent. Committed to growth. But sometimes committed to their own growth at the cost of others. The question was: how do you set the tone?"

According to Satya himself, as documented in the first few pages of his international bestselling book *Hit Refresh*, he stopped the session, looked at his team, and said: "Mike, I want to know the people in this room. I want us to know each other. I want us to drill down to the truth of who we are, and then to have the courage and the vulnerability to share it."

Those training sessions with Dr. Michael Gervais signaled a new era for Microsoft, and for businesses worldwide. And it was kickstarted not by a strategy announcement or product launch, but with a choice to invest in a new and different mindset.

But before we talk about the massive cultural transformation that Michael and Satya inspired, we need to be honest about what Microsoft had become.

From Know-it-All to Learn-it-All

When Satya Nadella became CEO of Microsoft in 2014, he inherited a company at war with itself. The tech giant, once renowned for innovation, had stalled out, crippled by deep infighting and hypercompetitive silos.

Employees were pitted against one another through a rigid performance review system. Executives had to rank every single person in stack order. The top 20% would get bonuses. The bottom 10% were transferred to new jobs... if they were lucky. Otherwise, they were booted out.

Not only that, but Microsoft had developed a defensive know-it-all culture. Many brilliant people approached projects as though they already had the answers. Arguing for your own solutions mattered more than asking questions. Being right was prized over being open.

I was working at Microsoft during this period, and I dreaded the review process. Each year, we executives had to gather with other team leaders in dreary conference rooms for days on end in order to analyze the performance of every single one of our thousands of employees, ranking them top to bottom.

From my own silo, I observed how the company had gathered incredible talent and extraordinary resources, yet allowed the culture to slowly strangle everything. And our partner organizations on the outside knew it. When internal teams are competing against each other, partners feel the friction. They stop bringing their real problems to the table because they realize they will not be heard.

In *Hit Refresh*, Satya describes the cultural transformation he drove when he took over, expressing his deep gratitude to Dr. Michael Gervais for helping craft the strategy. The prescription was radical in its simplicity: become *learn-it-alls* instead of *know-it-alls*. This approach meant being more curious, more open to changing course, and more interested in discovering new ideas than defending old ones. Hero culture was put to rest. Employees were no longer ranked against each other, but rewarded for collaborative work.

"Our industry does not respect tradition," Satya wrote. "What it respects is innovation." His new management style was built to encourage innovation.

When this cultural shift took hold, the consequences were visible to me across the partner ecosystem. For over 30 years,

Microsoft had guarded its software licensing model. We had a league of 50,000 field sellers, but lacked partners to co-sell with them. By now, we were three years behind in the cloud wars and desperately needed to catch up.

I served as part of the leadership team tasked with solving this problem. I'll tell the full story in Chapter 9. For now, I'll just say that the solution involved coordinating efforts across sales, licensing, legal, finance, and partner organizations.

The teams that embraced the new collaborative approach became the most effective in partner relationships because they showed genuine curiosity about what their partners needed. For them, it wasn't just about what Microsoft wanted to sell, but rather how two organizations could learn from each other and grow together. None of this could have happened in the know-it-all era.

The same dynamic plays out in every organization navigating the tectonic shifts we're experiencing today. The ones that thrive adopt a growth mindset that goes beyond a simple corporate initiative or offsite exercise. Their culture encourages curiosity and collaboration as a daily practice, and they build it one decision at a time.

Microsoft is not the only hyperscaler that treats mindset as an existential issue. Jeff Bezos fought the same war at Amazon, calling his initiative Day 1. "Day 2 is stasis," he warned shareholders, "followed by decline... followed by death."

Different vocabulary, same conviction: mindset starts at the top, with leadership actively promoting and rewarding it.

Why Mindset Comes First

Before we can talk about the rest of the partnership success principles highlighted in this book, we have to talk about mindset. Having the correct mindset lays the groundwork for executive commitment, shared vision, maniacal focus, and all the rest.

Begin by asking yourself this question: "Does my organization believe—genuinely, structurally, at every level—that working with and through partner organizations is a profitable and valuable way to achieve our goals? Not as a slogan, but as a lived belief?"

If the answer is *yes*, everything that follows is about execution. If the answer is *no*, or even *sometimes*, then no amount of strategy, investment, or program design will compensate for the mindset gap.

Mindset is foundational, which is why it comes first. Get it wrong and everything built on top of it will eventually collapse.

A Growth Mindset

The concept of a "growth mindset" comes from Stanford psychologist Dr. Carol Dweck's research, popularized in her book *Mindset: The New Psychology of Success.* Satya Nadella gave every Microsoft employee a copy of it when he became CEO. Reading it was not a recommendation, but a directive.

A growth mindset is characterized by the belief that your abilities, intelligence, and talents aren't fixed traits set in stone, but rather skills you can develop through effort, solid strategies, and learning from others. For example when you fail, you don't deny it or see it as a sign that you're utterly lacking in talent. Instead, you view the failure as feedback and an opportunity to improve moving forward.

Satya initially brought Dr. Michael Gervais into Microsoft to help implement a growth mindset at a practical level. When I interviewed him on the Ultimate Partner podcast, Michael framed the issue this way: "Mindset is simply a particular way you're approaching something. To have a growth mindset over a cup of tea with friends is easy. But what happens when stress enters the room? Under stress, we fray. We blame. We point fingers. We try to save ourselves. So you have to actually train your mind—the same way an athlete trains their body—to maintain the mindset that matters when it matters most."

Michael's key insight is that mindset is not a personality trait, but a training target. And it degrades if not actively maintained.

Having a growth mindset applies directly to successful partnerships. I often talk about it in my podcasts and at UP LIVE events. I also refer to it as an "infinite mindset," a term Simon Sinek popularized in his book *The Infinite Game*, a "partnership mindset," and an "abundance mindset." All these terms mean essentially the same thing.

The most common manifestation of a fixed mindset in partnership organizations is the scarcity mentality. Your people believe that sharing information, sharing customers, and sharing deals diminishes your position.

A growth mindset or abundance mentality, on the other hand, means the people in your organization genuinely believe that the whole is greater than the sum of its parts. Bringing a partner into a deal will create more value than going it alone. The ecosystem will multiply your reach rather than diluting your margins.

In the context of partnerships, your organization's growth mindset will be tested severely exactly when you need it most—when a deal goes sideways, a hyperscaler changes its priorities, a co-sell motion fails to deliver expected pipeline, or a competitor moves into your space. These moments truly reveal whether your mindset is real or merely decorative.

Consultant Jay McBain, who has studied the partner ecosystem deeply, loves to frame concepts quantitatively. He points out that partners who master the full customer journey around a hyperscaler platform can earn $3 to $7 for every dollar the hyperscaler earns—by extending, enhancing, and delivering on the promises the hyperscaler makes to the customer.

This is an abundance mindset in action. It is concrete evidence that the partnership opportunity is large enough for everyone to win. Your job is to capture your share by delivering real value, not to hoard access or protect territory.

The Frontier Firm in the Age of AI

In the age of AI, a growth mindset is the minimal entry requirement.

The pace of change is too fast for incremental adaptation. The organizations that thrive are pushing harder, deliberately operating at the leading edge. Ashleigh Vogstad of Transcends gave that posture a name when we sat together at one of our events, calling it "a frontier mindset".

Today's winners are asking how they can partner with Microsoft, AWS, Google Cloud, Anthropic, and OpenAI not to build massive data centers, but to build what runs on them. They're experimenting with agentic AI, developing proprietary solutions before they become standard, and treating their own internal transformation as a proof point to sell to customers.

Cyril Belikoff, Microsoft's VP of Marketplace, has developed a methodology for implementing this accelerated growth mindset in today's world: the Frontier Firm framework. It's a success model built on four pillars, which define what it means to be a leading organization in the age of AI.

Pillar 1: Enriching employee experiences. With this pillar, your organization examines how it is using AI to help its people do their work better. We're not just talking about developers and engineers, but also marketers, salespeople, finance teams, and HR.

"AI is changing our daily operations," Cyril says. "How can people get their heads around what AI is and then leverage AI more?" The Frontier Firm does not pilot AI in one corner of the business. It embeds AI everywhere.

Pillar 2: Reinventing customer engagement. This pillar means asking: How is your organization rethinking what customer interaction looks like in an AI-enabled world? Right now, a lot of companies are adding a chatbot to their existing motions. The companies that will dominate going forward go way beyond this. They're genuinely reimagining the touchpoints, timing, and value delivered at each phase of the journey.

Pillar 3: Reshaping business processes. This is where real organizational leverage lives. AI and agents can remove friction, accelerate decisions, and eliminate the kind of manual coordination that consumes enormous human energy, and yet *not* create proportionate value. As Marc Monday of ServiceNow put it, "Automating a bad process is still a bad process."

The frontier firm does not automate its existing workflows. It redesigns them first, then builds AI on top of what it has fixed.

Pillar 4: Bending the innovation curve. This forward-looking pillar looks at new experiences, new applications, and new business models that AI can potentially make possible for customers. What projects can you develop that hyperscalers cannot execute alone?

Partners making genuine progress across all four pillars—beginning with their own organizations—are the ones hyperscaler sellers will bring their most important accounts to first.

"Become customer zero for AI," recommended Nina Harding, Corporate Vice President for Americas Global Partner Solutions at Microsoft. "Because if you're not using it yourself, you can't credibly lead customers through the transformation."

The fixed mindset organization waits for proof—case studies, references, someone else to go first. The frontier firm becomes customer zero, making itself the proof.

Empathy: Practice, Not Principle

When he was working with Satya Nadella to implement a growth mindset at Microsoft, one of the most important qualities Dr. Michael Gervais emphasized was the role of empathy. He believes that empathy can and must be practiced and developed in order for organizations to thrive.

"A word on a wall doesn't change culture, and it doesn't change people," Michael said during our podcast conversation. "Empathy needs to be practiced. Step one is awareness. Mindfulness practices help you recognize whether you're being empathetic or defensive in a given moment. Step two is front-loading the training. Don't wait for a high-stress partnership moment to practice empathy. Train for it the same way an athlete trains—before the game, not during it."

For partnerships, this is profound. When two organizations come together, they'll most likely have different cultures, fiscal years, incentive structures, and leadership styles. Friction is inevitable. The organizations that navigate difficult moments with genuine empathy will build partnerships that survive these challenges and come out stronger. The ones that default to defensiveness and blame when pressures arise will have partnerships that die.

Microsoft embraced empathy under Satya's leadership so deeply that the company eventually shared their best practices with partners, helping partner organizations apply the same mindset training that had transformed their own internal culture.

"Partners were asking, 'How are you doing this change at scale?'" Michael said. "And one of the ways we did it was by helping people train their minds."

Failing Fast

One of the most practical expressions of a growth mindset is the willingness to fail fast. At Ultimate Partner, we often coach businesses to run pilot programs, make changes in small increments, learn rapidly, and iterate. This approach enables you to experiment before the cost of failure becomes catastrophic.

Organizations that approach partnership execution like a moon rocket launch—spending years on planning, making massive investments, and waiting for a single grand moment of activation—frequently fail. When the program does not work perfectly out of the gate, they often abandon it entirely.

The organizations that succeed treat partnership execution the way great teams treat product development. They're sprint-based, hypothesis-driven, and quick to adjust. They're not afraid to look back on a missed co-sell motion or marketplace campaign and ask, "What can we learn from our failure, and what do we try next?" Once they've tested and gained traction, then they build a flywheel.

Dr. Michael Gervais says part of a growth mindset in action is optimism, which he calls the center of mental toughness. "Optimism is the fundamental belief that things are going to work out if you stay in it longer and figure it out," he said. "I haven't met one world-class performer who is a pessimist. Not one."

A growth mindset, applied to partnership, ultimately requires the conviction that it is going to work, combined with the discipline to learn and adjust until it does.

The Last Freedom

Throughout my career, I have returned time and again to the seminal book *Man's Search for Meaning*. It was written by Viktor Frankl, a psychiatrist who survived the Holocaust and emerged from it with a mindset most people never tap into at all. He came away with a clear understanding of what it means to be human, and how to thrive in the most horrendous circumstances imaginable.

For years, Frankl and nearly all the people he loved, everyone in his community, lost everything. Those who weren't killed had their money, possessions and livelihoods, then their families, health and freedom taken away when they were forced into Nazi Concentration Camps. What he observed was that even in the most extreme conditions, one freedom remained.

"The last of one's freedoms is to choose one's attitude in any given set of circumstances," Frankl wrote.

To me, this represents a growth mindset distilled to its most essential form. You can't control what happens to you in life. No one is immune from disruption, loss, and failure. But you can control how you respond.

The pace of change in the partnership industry will not slow. Hyperscalers will constantly reset their priorities. AI will continue to rewrite the rules of markets you thought you understood. You can't control or even predict the shifts, but you can ensure that you and everyone in your organization adopts a growth mindset.

This is where the work of successful partnerships begins.

YOUR TURN

A growth mindset is foundational. Without it, every other principle is harder to build and easier to lose. These questions are designed to surface where the mindset work is most needed, in yourself and in your organization.

1. Think about the last time your partnership strategy did not deliver the expected results. What was your organization's instinctive response: blame the external environment, or ask what you could learn and do differently?

2. Who in your leadership team most consistently exhibits a growth mindset toward partnerships? What specifically do they do that makes this visible? And who is the greatest barrier?

3. Viktor Frankl wrote that the last freedom is to choose one's attitude. In the context of the tectonic shifts hitting our industry, what attitude is your organization choosing? Is it the attitude of an organization that will adapt and win?

YOUR COMMITMENT THIS WEEK:

Identify one area where your organization has been stuck in a fixed mindset about partnerships—a belief that has not changed even though the evidence suggests it should. Write it down. Then write what a growth mindset response to that same situation would look like.

CHAPTER 5

PRINCIPLE 2:
EXECUTIVE COMMITMENT

"We are fully committed to our partner strategy."

I have heard that promise, in some version, from almost every organization I have ever worked with or advised. The leadership team says it in the job interview. The CEO declares it in investor meetings and from the stage at the annual conference. The CRO writes it into the strategy deck and repeats it at the quarterly reviews. The CMO swears the partner marketing budget is coming.

And then nothing changes.

The compensation plan still rewards direct selling. The CRO guards their numbers from partner involvement. The CMO spends nothing on partner marketing. The CFO funds a partner team of two at a company doing hundreds of millions in revenue.

I have also watched this cycle repeat, over and over again, in my own career. Newly-minted Channel Chiefs and Chief Partner Officers arrive at our UP LIVE events a few weeks into the role, still glowing and quoting their CEO's all-in speech.

Then the walls go up, one department at a time, and the clock starts ticking for the partner leader. Six months to frustration. Twelve to a quiet LinkedIn profile refresh. Eighteen to departure. Two events later, someone new is wearing the partner badge and telling the same hopeful version of the same old story.

I know their pain because I've lived in their shoes.

This is not a talent problem. It never is. It is an executive commitment problem. And it is the most preventable failure I have seen in four decades in this business. It has also never been more expensive.

Half-committed organizations no longer get half the results. They get passed over completely. In this era, commitment is what separates the winners from the losers.

This chapter is about the difference between saying yes and meaning it.

What Commitment Actually Means

Nina Harding, Microsoft's Corporate Vice President for Enterprise Partner Sales in the Americas, says, "Partners are now front and center. They are one of our top three priorities. We are bringing them shoulder to shoulder with our sales teams—not as a side motion, but as part of how we operate."

That is what genuine organizational commitment looks like from a hyperscaler's perspective. The question is whether your organization reflects that same level of commitment internally.

Commitment means that everyone in your organization—from the C-suite on down to the selling floor—speaks partner. The CEO preaches its importance. The CRO commits resources. The CMO aligns on messaging. And the selling organization sells with, through, and alongside partners.

When I walk into companies and test against these dimensions, which I do every time, I find that most are failing on a few. Leaders confuse talking about commitment with demonstrating it. They mistake the presence of a partner program for the presence of a partner culture. They believe that hiring a Channel Chief is the same as becoming a partner-first organization.

It is not.

Executive commitment to partnership means more than just hiring one person to lead the effort, approving a limited partnership budget, and then moving on. And it requires more than a strategy slide in a deck and an announcement at one of the largest partner events of the year.

It demands nothing less than a complete restructuring of how the company allocates resources, compensates people, measures results, and holds its leaders accountable.

The CRO integrates partnership into the core sales motion—not as a parallel track, but as the primary go-to-market engine. The CMO embeds partner GTM into every demand generation program, campaign, and content strategy. The CFO funds the resources, technology, and compensation structures that make co-sell possible. Customer success and professional services delivery are structured to include partners by design. And the CEO actively champions the partner strategy in every executive conversation, board review, and all-hands meeting.

Anything less is simply a platitude.

CHAPTER 5

The Tony Safoian Standard

I've been lucky enough to witness the gold standard of executive commitment to partnerships up close.

When I was leading the U.S. Public Sector partner business at Microsoft, we were looking for early cloud partners—the ones who saw what the platform was becoming before the market did. I was introduced to Tony Safoian, the CEO of SADA, and flew out to Los Angeles to have dinner with Tony and his father Hovig, who had co-founded the company.

I expected to meet a partner pitching for position. Instead, I met a family that had built a partner organization on a foundation of trust and commitment. It was obvious that Tony understood exactly what the hyperscaler relationship could become if you treated it seriously enough.

As I worked closely with SADA in the years that followed, I watched how Tony led. He would not hesitate to hop on a plane and fly out to meet Microsoft executives at a moment's notice if it mattered to our work. He appeared on stage at events, which was uncommon—usually the partner leader would speak, not the CEO. He took great interest in the teams, jumping on calls he didn't need to be part of, and showing up for people in incredible ways. The level of executive commitment he consistently brought was unlike just about anything I've ever seen across the partner ecosystem.

And then in early 2019, Tony called me to share a decision that surprised most people: he was going to sell off SADA's Microsoft practice. He saw an opportunity to become Google's largest partner. At Microsoft, SADA was one of many partners. At Google, SADA could become the defining partner for an entire hyperscaler relationship. He was going all-in.

"Everything I've done for the last 23 years has been about creating best-in-class, best-in-breed," he told me. "I've been very focused on tilting the playing field, taking every opportunity we could grab. Because who wants to compete on an even playing field? Speciality, niche, scale, customer incumbency, being on the forefront—that's so much more fun."

"If you own the whole value chain, you want to implement as efficiently as possible so you can point to that value," Tony continued. "And when the next project and the next workload come around, you want to be there by earning the right to be there over and over again."

That last sentence epitomizes what Tony is all about. Not winning the deal, but earning the right to stay. And his deep executive commitment to partnership has certainly paid off.

SADA became Google's single largest reseller globally, and crossed a billion dollars in annual bookings. The organization was recognized as a Google Cloud Partner of the Year across multiple different categories every year for eight consecutive years from 2018 to 2025.

Then came the next big win.

Insight Enterprises, a Fortune 500 global solutions integrator, decided it was time to make a seismic move into the multi-cloud space. They wanted to acquire a top-tier company to accelerate the process. So Dee Burger, the President of Insight North America, called a senior Google executive and asked, "Which of your partners should I be talking to?"

It took three seconds for the Google executive to respond with a single name: SADA.

Dee and Tony agreed to meet in Scottsdale. The culture fit, and the deal closed on December 1, 2023.

There was no lengthy selection process. No one at Insight even created a shortlist. Dee simply trusted Tony's 23 years of commitment to one partnership, and the hyperscaler's enthusiastic endorsement. SADA had built a reputation so clear it needed no vetting. Tony has since retired.

I have watched a lot of partnership careers. Tony's was different—not because of the numbers, though SADA's numbers were extraordinary. But because of his conviction.

That is what executive commitment looks like when it builds for over two decades.

The CPO Problem Nobody Is Talking About

An increasingly common solution these days to building organizational commitment is appointing a Chief Partner Officer.

Janet Schijns, a former Verizon and Motorola channel leader, makes the case that the CPO is becoming one of the most important executive roles in technology today. "The CPO is about managing up, sideways, and down—creating something so exquisite for customers through partnerships that it cannot be replicated," she said. "And that requires a trust ecosystem where partners are treated as well as your own employees."

She is right.

In order for a CPO to succeed, the role must be clear and the individual must have CEO-level support and visibility. Without this, they'll fight an uphill battle every single day.

Greg Sarafin, the Global Manager of the EY Partner Ecosystem, pointed out the challenge. "If I asked ten smart people what the role of a CPO is and what the career path looks like, I would get ten completely different answers. Maybe eleven—because one would change their mind." He paused. "If I asked those same ten people what the role of a CFO is, I'd get very similar answers. Both the role and the career path."

That gap tells you everything about where the Chief Partner Officer sits in most organizations today. The CFO role has a century of institutional clarity behind it, but the CPO role is still being invented. Not only that, but people are often handed the title as a way of organizations acknowledging the partner function without truly funding or empowering it.

"It's a glory title we gave because we thought it might raise our throughput production, but we don't really have CPOs," said Janet Schijns. "The CPO isn't actually a member of the C-suite."

The most successful organizations are closing this gap. And the ones that close it fastest—by giving their Chief Partner Officer genuine authority, compensation alignment, and board visibility—are the ones building the most durable ecosystem advantage right now.

Why the CEO Is Often the Most Effective CPO

Here is a counterintuitive idea I have come to believe deeply, in large part thanks to watching Tony Safoian in action as CEO of SADA: In many organizations, the CEO is better positioned to drive partner strategy than anyone with "partner" in their title.

While the CEO rarely has the deepest expertise in co-sell mechanics or marketplace operations, this is the only person in the organization with the authority to force partner integration across every function, from sales, marketing, finance, and customer success to delivery. They're the only person who can hold every executive accountable to partnership as a strategic priority.

When a CEO shows up in a partner QBR, when they include ecosystem metrics in board reporting, when they personally engage with hyperscaler executives at the leadership level, and when they tie compensation to partner outcomes, they create organizational permission that no Chief Partner Officer can replicate on their own.

Jay McBain has made this case forcefully. He points out that the new buyer is already here. The platform economy is rewriting the rules of every industry. The death of the third-party cookie is moving marketing from interruption to influence. AI is creating a $4 trillion opportunity that can only be captured through ecosystem. And the mid-market represents an incredible opportunity that partner-led organizations are uniquely positioned to seize.

"This is why the next CEO should have partnerships in their resume or be a partnership leader," Jay said.

Microsoft, he observed, did not beat AWS with a better price or a better product. "What they did is have better coverage of the seven people that surrounded every buyer. They had more Microsoft endorsements sitting at the table—in the ebooks, at the events, in the podcasts, in all the different moments the customer was going through. That's one example. But that's the future of selling—a surround strategy of seven trusted people, and how you get them to influence that buyer in those first 28 moments."

Executive commitment to partnership produces a strategy that makes your organization the most trusted option before the buyer even knows they are deciding.

The Executive Commitment Diagnostic

Time and again, I've seen what happens without executive commitment. It's simple: partnerships fail.

The good news is, commitment is more than just a philosophy. It is a set of observable behaviors that you can test in your organization right now.

Before engaging any organization through Ultimate Partner, I apply five diagnostic questions to assess their genuine level of executive commitment. I share them here because I find they are equally useful as a self-assessment.

1. **Where does the partner function sit on the org chart?** A Chief Partner Officer who reports directly to the CEO or has genuine proxy authority from the CEO signals real commitment. A Channel VP buried three levels under the CRO is a structural signal that partnership is not a priority, regardless of what the org chart notes say.

2. **How is the field compensated for co-sell?** If field sellers receive no incremental benefit for engaging partners in their deals, or if they are measured in ways that make partner involvement feel like a threat to their numbers, this means the compensation structure is undermining the partnership strategy every single day. No strategy document will make up for this. Compensation must be fixed.

3. **What metrics does the CEO review for partner performance?** If the CEO only hears about partnerships at the annual partner summit, that tells you everything. The partner-sourced pipeline, partner-attached revenue, marketplace transactions, and co-sell metrics should show up in regular business reviews.

4. **When did the CEO last engage with a hyperscaler executive at the strategic level?** I'm not talking about a handshake at a conference. I'm asking when the CEO last had a substantive conversation about joint business planning, customer outcomes, and mutual investment. The answer should be clear and recent.

5. **What happens when direct sales and partner interests conflict?** This moment reveals more than any strategy document ever will. Does the organization default to protecting direct revenue, or to finding the answer that best serves the customer

> and builds the ecosystem? The resolution to that conflict, repeated hundreds of times a year across the field organization, shapes the partner culture far more than any executive proclamation.

My advice? Explore these five questions before you take a partner role, not after.

The Hardest Lesson

I learned this lesson the hard way right around the time Covid hit. When I took the role of VP of Channel and Ecosystems at Salience, I thought I had executive commitment behind me.

I did not.

The CRO stood staunchly in my way, refusing to give me access to the resources I requested. The CMO devoted all their efforts to direct sales. The sellers treated partner-led implementation as a threat, not a lever. And the CEO, who had spoken so compellingly about ecosystem-led growth in every one of our pre-hire conversations, showed little interest in partnerships once I took the job.

I made progress. I built relationships, moved metrics, and helped the organization see value in the Microsoft co-sell motion. But I could not move the culture. Culture does not change through partner strategy. It changes when the CEO changes what gets funded, what gets compensated, and what gets rewarded. That willingness was never there.

So let me say it again: do the diagnostic before you take the role, not after.

Building Commitment Where It Does Not Yet Exist

If you are reading this chapter and recognizing gaps in your organization, if you know you're not yet operating with the level of executive commitment that partnership success requires, then begin with an honest conversation. And bring the data.

Ninety percent of enterprise technology deals today involve partners. The organizations that have achieved pinnacle status with Microsoft, AWS, and Google Cloud, including Elastic, SADA, AvePoint, and others, did so because of organizational commitment. And as mentioned earlier, partnering done right pays: $3 to $7 for every hyperscaler dollar.

In watching this play out over the years, I've learned that commitment is caught more often than it is taught. Even a great presentation rarely converts a skeptical CEO. A room full of peers does.

So change the room.

Bring your CEO or CRO to one of our Ultimate Partner LIVE events. Let them sit among leaders who have already made the decision. Let them hear a hyperscaler executive say, "Partners are front and center." Let them listen to the CEO of a company larger than yours explain why they went all-in. When your chief executive hears it from another chief executive, the case stops being yours alone to make.

And if you cannot get them on a plane yet, put this book in their hands. I wrote it for you and I also wrote it for them. It can help convince executives how much saying no costs, and show them what saying yes requires.

The case for executive commitment is not hard to make. What is hard is creating the conditions for authentic commitment—the kind that shows up in compensation structures, organizational design, executive behavior, and resource allocation. Your company's partner success depends on it.

YOUR TURN

Executive commitment has the power to unlock your partnership strategy. These questions will help guide your way.

1. *Run the five diagnostic questions from this chapter on your organization right now. Where does the partner function sit on the org chart? How is the field compensated for co-sell? What metrics does the CEO review for partner performance? When did the CEO last engage a hyperscaler executive at the strategic level? What happens when direct sales and partner interests conflict? Then consider: what do your answers tell you?*
2. *The average Channel Chief tenure is under two years. If you are in that role or responsible for partnerships, what structural commitments does your organization need to make to ensure this is different? If you lead the organization, what would a genuinely committed partnership structure look like?*
3. *Tony Safoian committed 23 years to a single hyperscaler relationship. What is the longest your organization has sustained a deep, committed partnership? What made it possible, or what caused it to erode?*

YOUR COMMITMENT THIS WEEK:

Identify one place in your organization where a compensation structure, reporting line, or resource allocation is working against your partnership strategy. Bring it to the right executive with a specific recommendation for how to fix it.

CHAPTER

6

PRINCIPLE 3:

VISION AND VALUE ALIGNMENT

In 2001, I was asked to build a government channel from scratch at Itronix, a company that Golden Gate Capital had just carved out of Chapter 11 bankruptcy. The problem was, our largest competitor owned the government market. We had no GSA schedule, established partners, or proven motion.

What we did have was a great product—a rugged computer built for conditions where commercial hardware failed. We also had a conviction that the right partners could help us get our product in the hands of the people who needed it most.

Then September 11 happened.

That morning, everything stopped. An Itronix colleague called me from Spokane, Washington and told me to turn on my TV. The second Twin Tower fell, covering New York City in suffocating ash. Witnessing the looks of horror on people's faces, and learning about the thousands of firefighters and citizens who had lost their lives, most of us couldn't help but feel a sense of despair.

But the stillness and helplessness didn't last long. Just hours after the attacks, I watched as bipartisan members of Congress gathered on the stairs of the U.S. Capitol to sing "God Bless America." Our nation sprang into action.

So did our team. We saw an opportunity to aid in the recovery. Immediately, we offered to send one of our young techies to Ground Zero with Itronix laptop computers, which had been created for exactly this kind of scenario. And as the truth surfaced about the acts of terrorism, our government began looking for partners in fighting a whole new war.

Within days, I found myself on Capitol Hill, taking meetings with Senators and Representatives, as well as leading defense contractors. It was a thrilling moment. All of us were joined together by our shared vision, inspired to solve a problem we didn't yet fully understand: improving national defense and homeland security.

The plane crashes in New York City and Washington, D.C. had revealed just how vulnerable America was to attack. The questions being asked in every room were: "What technology exists that can help? Who has it? And how do we get it deployed fast?" Helping our country was our north star. Regardless of political party affiliation, industry or role, we were intent upon keeping our citizens safe.

Itronix had an answer. We knew our existing computers could help the military do their work better. But implementing that solution required partners—system integrators, defense contractors, and government allies who believed in us. Not a joint business plan. Not a co-marketing agreement. But, as Dr. Michael Gervais put it, a genuine shared conviction that together we could build something that did not yet exist.

Another clear need also emerged directly from these conversations. Itronix quickly began developing a brand-new technology: a wearable laptop designed for soldiers and first responders with a heads-up display that was ruggedized and field-deployable. Once we had aligned around this concept, we partnered with General Dynamics and took it to Capitol Hill. The federal government funded it.

Itronix. General Dynamics. The federal government. Three organizations with very different missions, cultures, and incentive structures that shared a single, focused vision.

Our partnership went far beyond a slide deck or even a series of strategy sessions. We were people who had decided together what we could build, and who were willing to commit our organizations to making this imagined future a reality. We stopped asking what each of us needed or what we'd get out of it, and started asking what the mission required. We'd done enough internal work to know what we stood for, and we found in each other a conviction that matched our own.

I have never forgotten what that felt like to be in those rooms on Capitol Hill in the fall of 2001. And I have been striving for that same level of vision and value alignment in every partnership since.

Finding Your North Star

We've established that a growth mindset and executive commitment are the internal foundations of successful partnership. Vision and value alignment is where the internal work meets the external reality. It's where two organizations decide on their north star—what they truly want to build together.

And it is where many partnerships go wrong.

Organizations frequently mistake alignment on language for alignment on substance. Two leadership teams walk out of a planning session having agreed on words. "We're going to drive AI outcomes together in the financial services sector," they say, without actually defining what this means in concrete terms.

They should be asking: "What customers? What outcomes? What does success look like in six months? Who is accountable for what?"

But in order to align with a partner, you first have to get clear on your own organizational values. This is why vision is the third principle of Internal Victory. You must complete the work for yourself before you can successfully execute any external partnership strategy.

Everyone wants to be the partner, like SADA, that field sellers navigate toward without prompting, the first one who comes to mind when the right account surfaces. Vision and value alignment is what creates that signal. It goes beyond strategic alignment to emotional alignment. When both organizations move beyond just agreeing that the direction sounds reasonable to genuinely believing in where they are going together, the result is enduring success.

In one of our deepest conversations on the Ultimate Partner podcast, Dr. Michael Gervais helped clarify what this work entails. "A philosophy is your first principles—the unwavering commitments that line up your thoughts, your words, and your actions," he said. "A vision is when you use your imagination to see something that has yet to take place, that you believe is so compelling you will do whatever it takes to make it a reality. A shared vision is when two or more people are using their imagination to see a state that has not yet happened, but is so compelling that they are motivated to work toward it together."

Internal first, then external. This is the sequence Michael insists on, and that my experience has confirmed every time. "You can't just borrow my vision and philosophy and make it yours," he explained. "You have to do the work for you to have yours, so that you can authentically say yes or no to a shared vision."

Most organizations skip this critical step. They rush to build a joint business plan without doing the internal work of understanding their own strategic intent and whether the partnership will truly serve it. As a result, they hit the same wall: stated vision and actual behavior diverge due to lack of conviction. The executive team goes in a direction the field does not yet believe in. The result is a partnership that looks aligned on paper yet feels misaligned in execution because the people doing the work are not all pulling toward the same goal.

The sequence is specific: determine your own north star, and articulate it clearly enough to rally your own organization around it. Then, and only then, bring your partner on board.

When Vision Alignment Leads to Success

When Ernst & Young handed Greg Sarafin a promotion to Global Managing Partner of the EY Partner Ecosystem in 2017, it hit him like a punch to the gut. "I was like, oh my God, my career is over," he told me.

Greg, an engineer, had come to EY from IBM. Two years into running a large piece of their financial services business for the Americas, he was a lead revenue producer with ambitions of climbing to the top of the corporate ladder. Instead, he felt like he was being pushed off the ladder entirely.

Many people I've talked with over the years feel this way. They view partner leader roles as a shove out the door because you're so unlikely to succeed.

Well, this was not the case for Greg. He went on to build one of the most successful partnership functions in the world, refusing to accept defeat. He approached the role the way an engineer approaches any system: by examining how things connect. His brain was wired to work backwards from outcomes, not forward from products, so that's what he did.

"Most organizations work their go-to-market from the inside out," he said during our podcast interview. "We've got all these products and services we need to sell, so we say, 'Let's figure out which clients to target.' I wondered, what if you flipped it? What if you said, 'What do my clients actually need?'"

That flip was foundational to his success. By asking what clients needed, Greg found his north star. He found he could determine which partner combinations would deliver outcomes better than anything either organization could build alone.

Greg also inherited a constraint that most ecosystem leaders would see as a liability. Because he was working at a Big 4 firm, certain hyperscaler relationships were simply off the table. Only Microsoft, SAP, ServiceNow, and Dell remained.

The good news was, these companies were all category leaders with significant enterprise penetration and aligned client outcomes. Fewer partners with deeper commitment meant clearer values alignment, which turned out to be an advantage.

"Many people think more partners equals more revenue," Greg said. "That is completely the opposite of the truth. Fewer strategic partners equals more revenue. Everybody else should be on your platform."

Greg's results speak for themselves. Over $10 billion in partnership-influenced revenue. 35% CAGR. 45% of total firm growth attributed to the alliance ecosystem. All built for under a billion dollars, on a constrained palette that most executives would have complained about, by a man who thought his career was over when he took the job.

What Greg did at EY with vision and value alignment, by focusing on client needs and building a few quality partnerships, lies at the heart of the third principle.

The One Thing

Partners, in my experience, tend to want to be all things to all hyperscalers. Yet even the clearest shared vision can become scattered in execution if organizations try to pursue too many objectives at once. That's why in my advisory work, I recommend partners focus on One Thing that becomes essential to who you are and how you do business. It differentiates you from the competition and helps you realize success.

In every high-impact partnership, there is usually one initiative—one customer problem, one market opportunity, one joint solution—that serves as the catalyst for the entire relationship. This key issue proves the partnership is real and creates momentum, attracting more resources, attention, and opportunity.

Finding the One Thing your organization wants to deliver requires asking three questions:

1. **What is the urgent, high-impact problem your partnership is uniquely positioned to solve?** I'm not talking about a general industry trend, but a specific problem that keeps your joint customers up at night.
2. **Does your One Thing directly support the hyperscaler leadership team's public commitments?** Whether you're working with Microsoft, Google, AWS, or OpenAI, does your initiative align to their stated solution area priorities? Alignment here earns executive sponsorship and resource support that no amount of relationship management can substitute for.

3. **Does it move the revenue needle with clear, measurable impact?** If the answer is no, it is a distraction, not a priority.

When both organizations can point to the same specific objective and say, "This is what we are building together," the partnership starts to compound.

The Bridge from Internal to External

Vision and value alignment helps guide your efforts with every principle that follows. A growth mindset opens the door to a successful partnership. Executive commitment funds the motion. Shared vision gives it direction.

But this principle is different from the first two in an essential way: it cannot be completed alone. The moment you articulate a vision, you must align on value with your partner. You step across the threshold from focusing on what's going on within your organization to how you can most effectively work with another.

This is when you lock arms.

Without all three Internal Victory principles working together, the Partnership Victory lacks a solid foundation. Now, with your partner standing beside you, the external work begins.

YOUR TURN

Vision and value alignment is the bridge between internal conviction and external partnership. Without a shared north star, even the most committed organizations will drift. These questions help you test whether your vision is real or aspirational.

1. *Can you articulate your organization's partnership vision in one clear sentence, in a way that a hyperscaler field seller would immediately understand and find compelling? Write it down. If you cannot, start here.*

2. *Think about your most important current hyperscaler partnership. Can both organizations tell the exact same story about why your relationship exists and what you are building together? If not, where does the story diverge?*

3. *Go back to Greg Sarafin's reframe. Most organizations work from the inside out, saying, "We have products, let's find customers." What would it look like to flip that completely in your organization? To start with what your customers actually need and build your partnership strategy around that outcome?*

YOUR COMMITMENT THIS WEEK:

Schedule a 30-minute conversation with your most important hyperscaler partner contact. The only agenda item: what does success look like for both of us in the next 12 months? Listen more than you speak..

CHAPTER 7

PRINCIPLE 4:
MANIACAL FOCUS

When I was a GM at Microsoft, the Worldwide Partner Conference (WPC) was the event of the year.

Each July, thousands of partners would descend upon the host city. Whether that was Toronto, Washington, Houston, or Los Angeles, the atmosphere was always electric. Microsoft made its biggest announcements about new programs, incentives, and priorities.

For me, personally, WPC meant back-to-back meetings from morning until night. CEOs and executive teams from each of our top partner organizations would come in, one after another. Most had prepared for this moment for months. They gave elaborate presentations, shared detailed plans, made revenue projections, offered solution area alignment and joint go-to-market roadmaps.

We would go through the presentations together, get genuinely excited about what we were going to build, align on the strategy, commit to actions, and walk out of the room fired up, giving high fives and handshakes. "This is the year we crack it," we'd say.

Then we all went home.

Two or three months later, I'd be in a meeting with my team of Partner Development Managers, going around the table asking for updates on the WPC meetings. "Whatever happened to that plan with Partner X?" I'd ask.

Crickets.

Too often, I'd discover there had been no follow-up. The strategy we had all been so excited about in July had never been implemented. The elaborate presentation was just sitting in a shared drive, opened occasionally but never operationalized.

And come next July, we would do it all again. Another great meeting. Another great plan. Another year when we never went from PowerPoint to execution.

But there was one exception.

One partner, Catapult, showed up every year at WPC with a book. Emily Lynch would create this record—a carefully organized account of every commitment the company had made to Microsoft the previous year, with evidence of exactly how they had delivered against each one. Then they'd walk us through it.

"This is your Scorecard," they'd say. "Here is how we achieved our goals together." I was always fully engaged because Catapult showed up with proof.

While many, if not most, of our partners focused on building a better slide deck, this company built a better system. That system, run consistently year after year, made them one of the most trusted and engaged co-sell partners in the entire Microsoft ecosystem.

This story demonstrates exactly what the fourth principle, maniacal focus, is all about. Clear goals with measurable outcomes. Weekly team meetings. Monthly executive meetings. An established rhythm. And absolute accountability.

CHAPTER 7

The OKR Framework

Once your organization is aligned with your partner on vision and value, the next step is implementing functional accountability. Without it, your vision is just inspiration.

Rebecca Jones, Chief Growth Officer at Bridge Partners, has spent her career watching partnerships form and dissolve, and she recognizes that the pattern is almost always the same. Two teams get in a room. The energy is real. The alignment feels genuine. And then everyone goes back to work and nothing changes.

"The partnership had vision," she said. "It just didn't have structure."

The organizations that successfully execute on a shared vision do something different. They translate it into specific, measurable commitments at every layer of the partnership.

The most effective tool I have seen for measuring success and maintaining maniacal focus is the Objectives and Key Results framework, or OKRs. I was introduced to the concept years ago, I can't remember exactly when. I've taken it into the room with me ever since, in partner meetings with every member of the organization from top to bottom, and across every function from product to sales and marketing.

OKRs, short for "Objectives and Key Results," serve as a popular goal-setting framework, which helps teams and organizations turn ambitious targets into actionable, measurable steps. The framework breaks down into two core components: the Objective, which defines a bold, inspiring, and qualitative target of where you want to go, and the Key Results, which are the two to five quantitative metrics used to track progress and prove whether you actually achieved that objective.

When applied to partnership, OKRs create a shared language for measuring progress across every layer of both organizations. At the executive level, they ask: what markets are we pursuing together, and what is each organization committing to deliver? At the go-to-market level: which customer segments are we targeting, who owns demand generation, and how are we amplifying a unified value proposition? At the co-sell level: which accounts are we jointly pursuing, how are field teams coordinated, and how is success measured and attributed? At the revenue level: what are the financial targets, how frequently are they reviewed, and who is accountable when they are not being met?

These are not aspirational goals. They are concrete commitments with named owners, aggressive timelines, and clear success metrics. Partnerships that are progress-focused and ruthlessly accountable will build momentum. Others will stall.

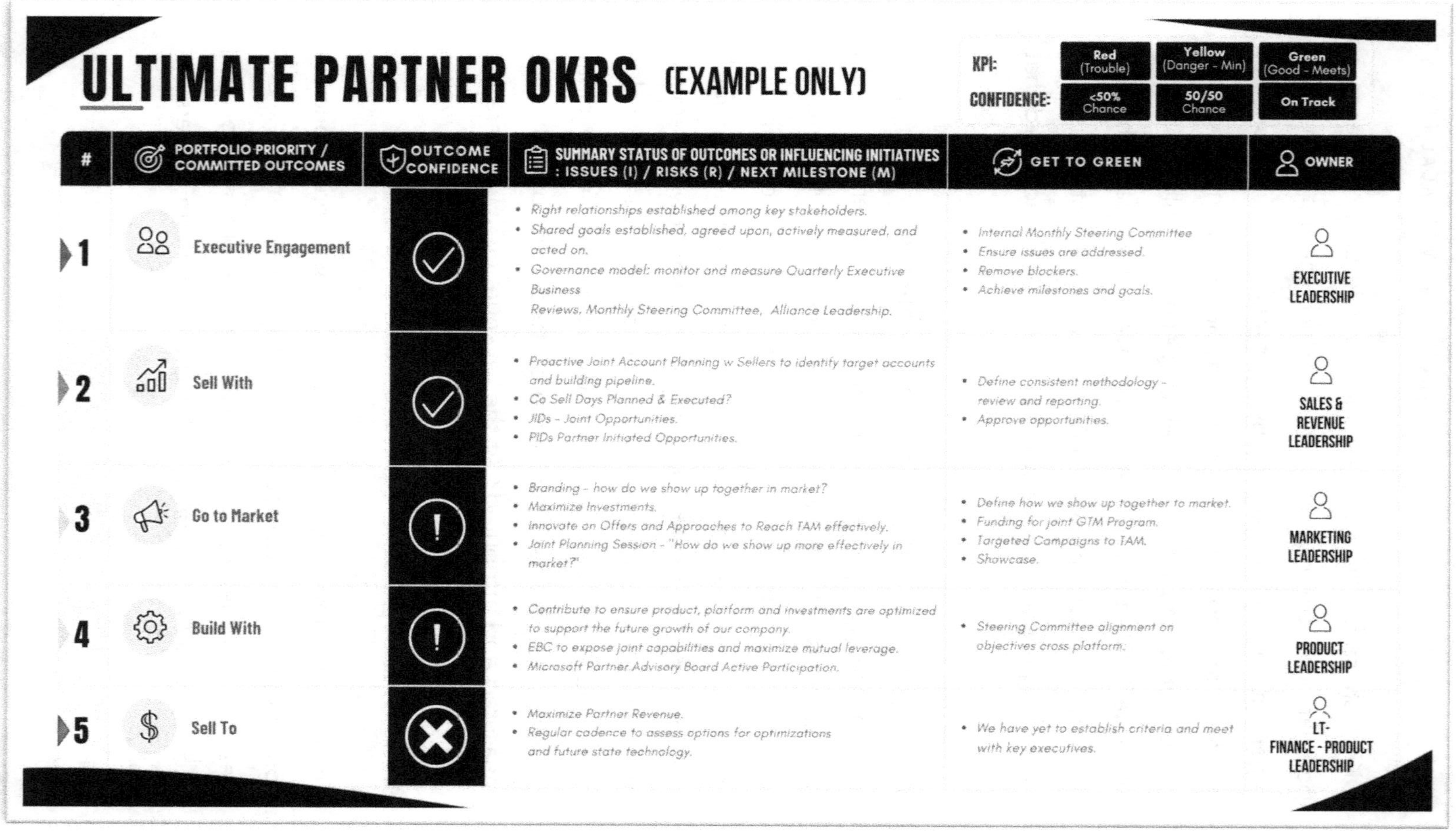

Figure 2. A sample OKR chart for measuring your partnerships.

The Steering Committee

Vision and value alignment combined with OKRs are not enough on their own, however, to ensure success. You need to build an infrastructure to hold it all together.

The Steering Committee ensures a high-performing partnership. This is when you take your OKRs, assign them to people, and then, in regular group meetings, hold each individual accountable for their performance. You rate each OKR as green, yellow, or red, with green meaning the job has been done, yellow indicating it is a work in progress, and red showing someone has dropped the ball.

In these high-accountability sessions, the standing question is not "How are things going?" but, "Are we green, yellow or red? If we are red, what is the specific action, owned by a specific person, with a specific deadline, required to get back on track?"

The best Steering Committees operate with three non-negotiable characteristics. They are cross-functional and cross-company, including representatives from marketing, sales, product, and delivery from both organizations. They escalate issues that surface in one meeting quickly, not allowing them to linger until the next. And they measure outcomes rather than activity—pipeline influenced, deals registered, marketplace transactions, customer results, etc.

When the Steering Committee operates this way, maniacal focus stops being an individual heroic effort and becomes an organizational capability. That is when results really begin to show.

What the Best Do Differently

At an UP LIVE event at Microsoft's Redmond campus, I sat down with Alyssa Fitzpatrick, Global Vice President of Partner Sales at Elastic and a former Microsoft executive. I wanted to understand, across hyperscalers, what separates the partners who consistently win Partner of the Year from the ones who never quite get there.

Her answer turned out to be one of the most operationally precise descriptions of maniacal focus I have heard. "Every partnership we run at Elastic, we look at what is the build piece. Where is the innovation coming from? What is the solution we are trying to build for our customers? Then how do we market that? How do we tell the world about what we built together? And then the sell. How do we empower our sellers to understand how to work together? You have to have that entire spectrum, and you have to decide what works for you."

"Build with. Go to market with. Sell with." Alyssa said that mantra had been in place at Elastic for nearly ten years. And if your organization is not running its hyperscaler partnerships through that same lens, then "I strongly recommend that you adopt it."

Beyond the mantra, what makes Elastic's model distinctive is the structural discipline behind it. Rather than having one partner leader trying to manage the entire Microsoft relationship, Elastic embeds dedicated points of contact from each functional area: engineering aligned with engineering, marketing aligned with marketing, sellers aligned with sellers. Each team talks to their counterpart team directly. Each team surfaces feedback through the right channel.

"If it is a build issue, you give it to the build team. If it is a go-to-market issue, take it to the marketing team. Landing the feedback in the right place means it can actually be addressed. Otherwise it floats around without resolution."

I call this "functional alignment." Alyssa calls it "swim lanes." The terminology matters less than the principle: for maniacal focus, you cannot have one person carrying the full weight of a hyperscaler relationship. The partnership has to be embedded across the organization—functionally, structurally, and systematically.

On the question of resourcing, Alyssa was honest about what maniacal focus actually requires. "When I came into Elastic, about 15% of the entire partner organization was dedicated to hyperscaler relationships. That was a conscious decision made before I arrived. It meant other partnerships received less attention. But that trade-off was understood, and it worked. Partners that do are the partners that win."

You cannot be maniacally focused on everything. The organizations that reach pinnacle status with a hyperscaler make the sometimes difficult decision to concentrate their resources on the relationship they most need to win.

I then talked about "the crickets problem"—how you have a great kick-off meeting, make an ambitious plan, but months later, silence follows.

Alyssa nodded immediately. "That is exactly right. And the remedy is what I call getting in the room. Get your CEO here. Get your CRO on a plane. We have meet with Microsoft twice a year—an entire day, executive to executive, engineering sessions that go into a room I am not even in. That commitment, showing up consistently at that level, is what makes the relationship real."

The bottom line, she said, is human relationships. "I have been in partnering for 35 years. And my kids ask me what I do. I tell them I build human relationships where they did not exist before. And then we create business out of that. That is what partnering is."

And that is the essence of maniacal focus.

Removing Obstacles

In every partnership, obstacles will appear. Projects get delayed. Compensation structures create internal conflict. A key executive sponsor moves on. The hyperscaler changes its fiscal priorities mid-year.

The difference between stalled partnerships and successful ones is maniacal focus. Because when you have maniacal focus, you remove blockers to achieve your OKRs with speed and decisiveness.

Reis Barrie, CEO of Carve Partners and one of the most rigorous practitioners of the Microsoft co-sell methodology I know, has spent years cataloguing exactly what causes partnerships to fail. The most critical error, he says, is lack of accountability.

"You have to have a dedicated partnership owner. Not someone who does partnerships alongside three other things. Someone whose job it is, full time, to manage the complexity and stay current as the programs evolve."

Compensation can also be a problem. "Misaligned sales compensation can doom a partnership before it even begins," Reis said. "If your field sellers are not incentivized for co-sell—or worse, if they are penalized for it by how their quota is structured—no partner program will fix that. It is a structural problem that requires a structural solution."

Another obstacle is committing to too many motions at once. Greg Sarafin said, "You only do big motions because the cost of sales is too high otherwise. My rule is, with any partner, what are the five motions that matter that are going to get you to 80% of your value creation revenue?"

Every motion carries a coordination cost before it returns a dollar. You cannot absorb that cost across 20 simultaneous efforts. The organizations that try end up with nothing working at the depth it requires.

The obstacles highlighted above—lack of clear ownership, compensation, and overcommitting—will prevent you from getting to green on your OKRs. Surface them. Assign ownership. And remove them. Every single week. A partnership that loses momentum is far harder to restart than one that never stopped.

Be disciplined. Maintain maniacal focus, and you will meet with success.

Go Slow to Go Fast

At a recent Ultimate Partner Winter Retreat, Erin Figer, founder of CORE Consulting, talked about how her methodology has produced 200% pipeline growth for partners. Our audience immediately began asking her about the tactics, campaigns, and tools she uses to generate such extraordinary results.

Her answer surprised them: "Go slow to go fast."

The counterintuitive truth about maniacal focus is that it does not start with speed. Before you operationalize, activate or co-sell a single deal, Erin said, you must slow down and build a solid foundation. Define your strategy. Agree on what success looks like. Understand how your solution drives the metrics that matter to every player involved—the partner, the hyperscaler seller, and the customer. Assign responsibility for OKRs.

"If you go slow to go fast," Erin said, "then operationalizing your co-sell strategy will happen quicker. The data will be more meaningful. You'll reach revenue faster, not slower."

To most organizations, this approach feels like a delay. It is not. It is the key to building a flywheel that starts spinning gradually, then becomes almost impossible to stop.

Bring the Book

Catapult always showed up to our annual Microsoft WPC meetings with a book. They did not have a better strategy, a better product or a bigger team. They had a system.

They directly aligned their measurement of outcomes to Microsoft's Scorecard, which was the accountability tool we used internally. And as a result, they stayed relevant to the field organization week after week, quarter after quarter, year after year. We spoke a common language.

Maniacal focus, done right, produces a compounding record of execution that makes you the partner a hyperscaler seller thinks of first, trusts most, and calls before anyone else. So build the system and then run it, year after year.

That is how you'll become an Ultimate Partner.

YOUR TURN

Maniacal focus turns strategy into execution. These questions are designed to surface whether your organization has a system in place or just a plan.

1. *Think about your last joint business planning session with a hyperscaler. What happened in the 90 days after that meeting? Were the commitments tracked? Were the blockers surfaced and removed? Or did the plan sit in a shared drive, unopened?*

2. *What is your organization's single most important partnership priority right now—the One Thing that, if executed well, would drive more co-sell momentum than anything else? Is that priority reflected in how your team spends its time and where resources are allocated?*

3. *Considering "go slow to go fast," where in your current partnership execution are you moving too fast? Activating too many motions before the foundation is ready, and creating noise instead of momentum?*

YOUR COMMITMENT THIS WEEK:

Take your current joint business plan with your most important hyperscaler partner and ask one question about every commitment on it: is this green, yellow, or red? For every red, name the specific blocker and the specific person responsible for removing it. This week.

CHAPTER

8

PRINCIPLE 5:

BRAND AND STORY

Early in 2025, the AvePoint team invited me to join them at the Nasdaq for the ringing of the closing bell. Minutes later, we spilled into Times Square, looked up, and saw our images on the giant screen. It was a powerful moment.

I had known the founders of AvePoint since the very beginning. And the beginning looked nothing like this.

Twenty-five years earlier, two engineers sat in a public library in Somerset, New Jersey, writing code. They didn't have an office. They didn't even have customers. All they had was a conviction that Microsoft's platform was going to matter, and that organizations would need serious help protecting and managing what they built on it.

Kai Gong and TJ Jiang were right.

From the inside, I watched their brand do its work. Soon, for any deal that touched data protection or governance, AvePoint was the first name in the room. I witnessed them earn their first Microsoft Partner of the Year award, followed by their second, third, fourth, and fifth. In time, they built a company with a Nasdaq listing and more than 28,000 customers in 100 countries, across more than 100 cloud marketplaces. Today, AvePoint stands for trust, helping organizations scale AI responsibly.

What's their secret? What have they done differently?

According to Dux Raymond Sy, AvePoint's Chief Transformation Officer, the company's success was built not on marketing campaigns, but on a brand story earned through decades of delivery. And Dux knows the power of that kind of story firsthand, because he built his own career on one, starting in an unusual place: a book.

Dux began his career in technical roles but quickly realized he did his best work not behind a keyboard but in front of people, translating complex ideas into concepts everyone could understand. So he wrote and published a practical guide, *SharePoint for Project Management,* which demonstrated his expertise before he ever walked into a meeting.

"I started my career as an entrepreneur," he told me. "I had to learn sales. I had to learn marketing with no money. And I thought, 'I'm gonna write a book to gain credibility and attention.'"

It worked.

When Dux joined AvePoint in 2013, the ecosystem was already moving toward cloud-scale collaboration. The shift created new opportunities and brought new risks. "Microsoft rounded up some key partners and said, 'We're going to do this. Who's in?'" Dux recalled. "And we raised our hand. 'Yeah, we'll build with you.'"

What followed was a long stretch of execution. AvePoint showed up time and again, aligning with the platform's direction and doing the work that enables progress at scale. They became widely recognized for helping organizations across Microsoft, Google, and Salesforce environments prevent overexposure and sprawl, modernize fragmented data, and recover quickly when things go wrong.

Through it all, Dux remained deliberate about one thing: AvePoint treated Microsoft like its best customer, investing in the relationship at every level. Whenever a decision touched the hyperscaler's space, someone in the room already knew the AvePoint brand.

That story still resonates today, through every shift our industry has experienced. In the field, AvePoint's brand carries a credibility sellers and customers can feel: confident yet humble, ambitious yet never aggressive.

I asked Dux what makes a brand that durable, especially inside hyperscaler ecosystems. The bottom line, he said, is trust. "I can give you the most Marketing Development Fund (MDF) or the most incentives and discounts. But if I can't trust you, it doesn't matter. It's a human relationship. I don't think that's ever going to go away."

That idea is more relevant than ever in the age of AI. Tools scale activity. Only trust scales commitment.

Your Brand Is the Body of Evidence

"Do I trust this partner to make me look good in front of my customers?"

This is the one question every hyperscaler field seller asks, consciously or not, when they consider bringing your organization into a deal. And it gets answered based not on your messaging, but on your reputation.

When we talk about "brand" in the context of hyperscaler partnerships, we're not talking about your capabilities, technical chops, company logo, positioning document, website, or social media presence—though all of those things matter. We're talking about the stories circulating about you among Microsoft, AWS, and Google Cloud sellers. Do you deliver on what you promise? What kind of value does working with you create for customers?

Think back to the Jupiter lighthouse I mentioned in the Introduction to this book. A lighthouse does not chase ships. It holds its position, burns consistently, and lets its signal do the work. The ships that already know the channel find their way without being summoned. The ones navigating uncertain waters are grateful it is there.

Your brand works the same way. You build a body of evidence over time, maintain it consistently, and offer the most value precisely when conditions get tricky. When a hyperscaler field seller faces a difficult account or a tight quarter, they reach for the partner they already trust, not the one with the slickest pitch.

The competition for trust is real. Microsoft has tens of thousands of partner organizations all competing for the attention of the same field sellers, who are measured on quarterly quotas of tens of millions of dollars. They gravitate naturally to the partners who have made themselves easy to understand, easy to position, and easy to trust.

Building a trustworthy brand is essential to becoming the ultimate partner.

The Story Hyperscalers Tell for You

Here is a standard most organizations do not apply but should: can a hyperscaler field seller, with no prompting and no materials, tell a compelling story about why they co-sell with you? Their story should be not about what your product does, but about a customer, a problem, and a measurable outcome.

Organizations that pass this test have done two things.

First, they have delivered results so specific and memorable that the stories circulate on their own. For example, it's not, "We help financial services companies with data governance," but "We helped a top-five bank reduce their compliance review cycle by 60% using our integration with Microsoft Purview, and we did it in under 90 days."

Second, the stories travel far and wide. They're told at events, in field briefings, in LinkedIn posts, in joint press releases, and in the partner success story program, giving sellers the evidence they need to advocate confidently.

When this happens, the impact on your organization is massive, because it means the hyperscalers are doing the work, sharing your stories and building your brand for you.

Nina Harding, a Corporate Vice President at Microsoft, described just such a program. Partners submit customer success stories branded with their organization's identity. Microsoft field sellers then use these stories directly in their customer conversations.

"Talk about credibility within the field," she said. "Your name, your recognition, are picked up and amplified by our marketing and our sellers."

You deliver results. You tell the story. Then Microsoft amplifies that story to the field. As a result, the field wants to co-sell with you. So you get more opportunities, you deliver more results, and business compounds. This is how the brand signal builds upon itself.

When Tony Safoian built SADA into Google Cloud Partner of the Year, year after year, he made sure that every customer win became a story. Not only that, but he ensured the story was told the same way every time—with precision, data, and better-together outcomes that validated customer experiences. Those stories are why sellers sought SADA out rather than waiting to be contacted, and SADA grew exponentially.

To be effective, the story has to circulate inside the ecosystem and outside it.

Building the Ultimate Partner Brand

When I launched the Ultimate Partner podcast in 2017, in a spare bedroom with nothing more than a USB microphone I had purchased online, Dux Raymond Sy was one of my earliest guests. We had to have the conversation twice because the first time, I forgot to hit the record button.

At the time, I had no audience, podcast studio, or brand. What I had was a point of view shaped by nine years at Microsoft, two decades of partnership experience, and a conviction that the leaders navigating the technology ecosystem deserved something better than what existed.

I built toward all that one episode at a time.

By the time Dux came back to the studio seven years and nearly 210 episodes later, the world had changed completely. The podcast had grown from a hobby into a full-blown production. We had built the Ultimate Partner community. We had become, according to multiple industry rankings, one of the most listened-to podcasts in the technology partnership space. And we had recently made a bold decision to take our brand even further.

By 2023, the world had reopened post-pandemic. Conferences were back. But Microsoft chose to keep their Inspire Conference as a virtual event for a third straight year. For thousands of partners, Inspire had been a vital gathering where you could get into the room with people who mattered, hear priorities directly from leaders, ask questions, and build relationships that drove your business for the rest of the year. Without it, there was a void.

I wondered if we could fill that void. A friend looked at me and said: "Just do it."

In 120 days, we built the first Ultimate Partner LIVE from nothing: concept to sold-out event. We secured Microsoft as the title sponsor. We filled a room in Las Colinas, Texas with 270 leaders. Corporate vice presidents flew in, as did presidents of multi-billion dollar organizations. Partners who had not been in the same room since the last live Inspire conference four years prior came together to talk about co-sell, marketplace, AI, and the future of the ecosystem.

I stood on stage and told them the truth. "I don't have all the answers. But I know who does. And that's why we've assembled this room."

To this day, I still receive feedback from the people who were there. It tells me something I sincerely believe: proximity, trust, and shared experience create a quality of conversation and a depth of relationship that no digital channel can replicate.

That first event was proof of concept: the Ultimate Partner podcast was not the brand; the community was. The podcast opened the door, but the live events brought our community together and told the story that made our brand a true success.

We are now in our fourth year of live events, having hosted 13 conferences across seven cities: Dallas, Reston, Redmond, Boca Raton, San Francisco, NYC, and Bellevue. Our community numbers in the thousands, and the podcast has crossed more than 300 episodes. Revenue has doubled year over year, and events have tripled in size.

None of this happened due to advertising, a PR campaign, or a marketing budget. It happened thanks to the same mechanism I have been describing throughout this chapter. We've consistently delivered a product that people find genuinely useful and, over time, have come to trust.

Today, when I stand in front of rooms full of leaders at our events and talk about what it takes to become the partner a seller reaches for first, I am not talking in theory. I have lived it. I built the brand that got me into those rooms, and I have watched our brand open doors that a resume or a cold outreach never could have.

Technology changes. Programs change. Platforms evolve. But the human being on the other side of the co-sell relationship—the Microsoft field seller, Google Cloud account executive, or AWS partner manager—still asks the same question before they commit to you. "Do I trust this partner?"

Their answer depends not on one meeting, but on watching you, over months and years, show up and deliver. You tell an outcome-based success story, they repeat it, and then you do it again.

Nine years ago, I had little more than a microphone and my conviction. Today, we are one of the most respected voices in the technology ecosystem. The brand did not precede the work. The brand was the work: accumulated, compounded, and made visible over time.

Start. Be consistent. Stay humble. Keep building.

YOUR TURN

Most organizations treat brand and story as a marketing function, when actually it is a partnership function. Answering these questions will help you assess whether your brand is working as hard as your team.

1. *Can a Microsoft, Google Cloud, or AWS field seller with no materials and no prompting tell a compelling story about why they co-sell with you? If not, what is missing? The customer outcomes, data, or consistency of the story you have been telling?*
2. *What is the one thing that makes you the partner a hyperscaler seller reaches for first in your solution area? If you do not have a clear answer, that is your brand gap.*
3. *Consider what AvePoint's 25-year arc demonstrates: a brand earned through consistent delivery rather than marketing campaigns, trust valued over incentives and MDF, and the deliberate practice of treating your most important platform like your best customer, at every level. How does your organization score on each of these three?*

YOUR COMMITMENT THIS WEEK:

Find one customer success story your organization has delivered in the last 12 months that has never been written up, published, or shared with a hyperscaler field team. Get it documented and into your co-sell assets this week

CHAPTER

9

PRINCIPLE 6:
DELIVERED OUTCOMES

During my last few years at Microsoft, we were as much as three years behind in the cloud wars. We still had a competitive edge no one else could match: 50,000 sellers in the field, calling on every enterprise in the world. The trouble was, we didn't have partners who knew how to sell with them.

The company was making the hardest pivot in its history. For 30 years, we had licensed software and sold products such as Windows Server and Exchange through OEMs. Now, in order to keep up, we had to learn to sell fully to the cloud. Office 365 and Azure served as our two entry points. But neither our sales motion nor our partner motion supported them.

We were selling a cloud future through machinery built for a licensing past.

In the Public Sector business where I was leading partnerships, the problem was even more acute. Our government customers were enamored with cloud offerings from other hyperscalers, but Microsoft was still only offering enterprise agreements.

I was asked to solve this conundrum. I knew that I had to get worldwide licensing, legal, and finance, as well as the partner organizations, to all move together. This was an immense feat. So I gathered the best and created a team.

Brian Campbell had worked with Bill Gates and Steve Ballmer back in the day. He knew Redmond cold—who to call, what would fly, where the bodies were buried. For the better part of a year, he led the effort to craft Microsoft's first licensing model built for the cloud: the Agreement for Online Services for Government, or AOS-G.

AOS-G became the precursor to the CSP program every partner knows today. All the earliest Office 365 licenses that were not coupled to an enterprise agreement flowed through it. And here is the part that still amazes me: more than 10 years on, hundreds of millions of dollars a year still flow through that model, much of it through Carahsoft, the largest reseller to the federal government. Brian deserves the credit for making it happen. My job was knowing he was the one who could.

But licensing merely opened the door. With Azure, we finally understood that our success would come from selling not just our own first-party software, but from third-party software running as tenants in our data centers. To spark that ecosystem, we needed to recruit ISVs, and most of them were already building on a competitor's platform.

What we had to offer was our edge: those 50,000 field sellers worldwide. What we had yet to build was the connective tissue: the relationships, account by account, seller by seller, that would turn an org chart into a co-sell motion. And this was before co-sell even had a name, let alone a program, a portal, and a playbook.

Not all of my teams cracked the problem at the same rate. The one that got there first was healthcare, led by Bill Hawkins. Bill's attention to detail and razor-sharp execution left nothing to chance. His success showed the rest of the organization what the motion was supposed to look like. Watching Bill work is where maniacal focus, the principle behind Chapter 7, first took shape in my thinking.

Only a handful of partners were effective in those days. Nasuni and 10th Magnitude, now part of Cognizant, became our go-tos, and growth followed. Two of the people in those early motions became friends of Ultimate Partner. Erin Figer was consulting with Nasuni, standing up their co-sell practice at the time. Jason Rook had left Microsoft to lead 10th Magnitude's growth as a partner.

We were all learning the same lesson at the same time: co-sell was not a program. It was a set of relationships, backed by proof, executed one deal at a time.

Erin took that lesson further than almost anyone. She went on to found CORE Consulting and built it into the go-to co-sell advisory for ISVs navigating Microsoft, AWS, and Google Cloud.

"Initially, people took co-sell to mean connecting a seller with a cloud seller, this seller-to-seller motion," she said. "Now people are thinking much bigger. When you build a solution and bring it to market, how are you going to co-sell holistically—with the marketplace, with the ecosystem you're part of, with all the partners who can support that solution?"

In just a few years, we turned our sellers and a handful of ready partners into a holistic, ecosystem-wide co-sell motion. And the cloud war we were losing when this story began? Microsoft caught up and then some.

That is the discipline this chapter is about: delivered outcomes. Every partnership eventually faces one question, and it often arrives faster than most people expect: what have we actually achieved together? A delivered outcome means accountability. In this ecosystem, it is the only currency that never devalues.

But there's another critical aspect to this principle, as well, and it's one that most organizations miss. Delivering the outcome is only half the work. You must also demonstrate it visibly, repeatedly, and in language the other side of the table actually values.

An outcome nobody can see might as well not exist.

What Outcomes Look Like

A decade after running those first motions at Microsoft, I sat on stage at our Ultimate Partner LIVE event in Reston, Virginia with Darryl Peek, who leads Partner Sales for the Public Sector at Elastic. He's part of a team that has achieved a rare feat: winning Partner of the Year recognition from Microsoft, Amazon Web Services, and Google Cloud, all at the same time.

When I asked Darryl what a healthy partnership looks like, I expected to hear him talk about relationships, shared vision, and alignment. Instead, he counted. "We measure the progression," he said. "Zero to one. One to five. Five to ten. Ten to 25. When we see that progression, that's where we're really starting to see health in the partnership."

What was he counting? Co-sell deals closed with a hyperscaler.

Darryl's answer has stayed with me because it mirrors how partner relationships actually grow. At zero to one, you are proving viability. At one to five, repeatability. At five to ten, you have a genuine motion. At ten to 25, a compounding one. Each stage demands more activity, more infrastructure, more trust, more marketplace maturity, and more customer proof points in order to create the next opportunity.

Darryl went on to reveal other numbers on his scorecard, as well, that were just as specific. Partner-sourced bookings and pipeline. Partner-attached numbers. Cloud consumption. Certifications. And perhaps the hardest yet most important one of all. "We track influence revenue. Not just how a partner is coming to us, but how a partner is helping us close deals they didn't originate," he said.

If you want to demonstrate your organization's value to a hyperscaler, speak to them in numbers that point to your delivered outcomes.

Yet Darryl also emphasized the importance of the human connection underlying it all. "The difference between contacts and contracts is the R. That's the relationship. Don't show up only when you need a renewal. Build the relationship. Tell them about your wins. Make them feel like a partner, not a transaction."

Proof You Can Point To

In Chapter 7, I talked about how, a decade ago, most organizations showed up at Microsoft's annual partner conference with presentations and promises. Catapult, on the other hand, showed up with a book. They walked us through a record of every commitment they'd made the previous year and showed us evidence of exactly how they had delivered against each one. That discipline was rare, and it made Catapult the partner we trusted most and chose to work with first.

Things have changed, however. These days, hyperscalers are building the book right into the system.

In July 2026, at the AWS Summit, Amazon announced that it would be giving partners an opportunity score. The following day, I sat down with Matt Yanchyshyn, VP of AWS Marketplace, to discuss what this shift means.

"People have been asking for this for years," he said. "So you can say, 'When we do this and we give AWS this information, the score goes up and we have a higher propensity to be co-sold.' Before, it was this mystery. You had to guess." Now, you can trace inbound leads through conversion to the AWS revenue your product drives, tie it to your score, and put that number in front of a skeptical field seller. You can say, "This is the business I am going to drive for you, and this is why my opportunities deserve your attention."

The Catapult book has become a dashboard. The proof is in the numbers. And the partners who build the discipline are the ones whose numbers speak for themselves.

For example, Matt told a story about DataMasque, a small company out of New Zealand. They had no field team or campaign, yet they landed 13 new logos through pure self-service on the marketplace. These deals included a Fortune 1,000 American enterprise that discovered DataMasque through AI search, and a customer in Peru, a country where they have never had any presence.

The warning Matt issued was blunt: partners without a self-service motion are "literally leaving money on the table."

The Money Most Partners Leave on the Table

For every dollar your hyperscaler earns in your accounts, how many dollars can you earn?

Canalys puts the answer, for great partners, at $3 to $7. IDC runs higher still: $8.45 in services and $10.93 in software for every dollar of cloud revenue. Cyril Belikoff, Microsoft's VP of Marketplace, puts Microsoft's own numbers at almost $9. And a software company, he says, can make almost $11.

When it comes to AI, the returns Cyril shared were even more striking. Customers are seeing four times ROI on average, and ten times for leaders. Not only that, but these returns land in just 14 months instead of the two to three years typical of major IT projects.

These figures reflect a moment in time, early in Microsoft's AI marketplace push. The trajectory has only accelerated since. The truth is, your opportunity is not measured in percentage points. It is measured in multiples of what the platform itself earns.

Yet most partners never see anything close to their multiple. They compete for a small slice—the resell margin, referral fee, or one-time engagement—and leave the rest on the table. Why? Because the dollars do not live in the transaction. They live in the work that wraps the platform. Consulting and advisory. Architecture. Implementation. Integration. Managed services. Optimization. Customer success. The work that compounds the customer relationship, because every delivered outcome creates the next opportunity.

And here is what should keep you up at night: the multiplier work is being won right now, in accounts you already serve, months before any deal hits a pipeline.

Jen Odess, Group Vice President of Partner Excellence at ServiceNow, says that 87% of all their implementations were executed by the partner ecosystem.

Not 30 percent. Not even 50. But 87.

ServiceNow's Now Assist AI is the fastest-growing product in the company's history. It has reached $500 million in annual contract value. That growth is not happening despite the partner ecosystem, but because of it. The winning partners in that world understand their role, which is not to sell the platform, but to make the platform deliver outcomes ServiceNow cannot deliver alone.

That shift from selling to delivering is the difference between transactional results and compounding ones.

The 28 Moments Framework

According to Jay McBain, the average enterprise purchase involves 28 distinct moments of research and influence made before any money moves. The most decisive ones are partnership moments—paid consulting, architecture, and design work that occurs months before a dollar is spent.

Jay analyzed a multi-million dollar SAP deal at AstraZeneca where NTT and SoftwareOne shaped the outcome in December. The vendors found out about the deal in April, four months after the money had been committed.

"They lost this deal without knowing there was a deal," he said.

Figure 3. Jay McBain's 28 Moments Framework. This shows exactly where work is won, and why most vendors never see it happening. Courtesy of Omdia

Here is what happened.

The customer was AstraZeneca, a company with history—ICI Americas, AstraZeneca's precursor, was the first SAP customer in North America. Months before a dollar moved, the deal was already happening. In December, partners came into the account with five different paid projects, including consulting, advisory, and architecture design.

SoftwareOne won by bringing deep Microsoft expertise. NTT, who had spent decades building a relationship with AstraZeneca, won three projects in August. Together, they shaped exactly what the customer would buy—before the customer downloaded an ebook and before a single marketing lead fired.

AWS got two direct visits to their website from AstraZeneca during that journey, but lost the deal. Microsoft won millions of dollars without a single salesperson ever working on the deal. Partners brought Microsoft in to solve problems AstraZeneca was having, and contracts followed.

The partners who won were not running better campaigns. They were already in the account, with a track record of delivered outcomes, when the 28 moments that mattered arrived.

The good news is, the technology needed to see those moments now exists. Cassandra Gholston, CEO of PartnerTap, put it this way: "Companies are moving from random acts of co-sell—everyone sending spreadsheets around trying to do top-of-funnel activity—to automation where you can securely share account and opportunity data, align field teams, drive white space plays, and orchestrate multi-partner engagements ahead of time."

That is what a delivered outcomes system runs on. A proactive, data-driven motion that surfaces the invisible pipeline before it becomes someone else's deal.

Your Outcomes Operating System

So what does it take to build an outcomes-based system, not just an outcomes intention? From the highest performers across hundreds of conversations, I've drawn five disciplines that work.

1. **Define what outcomes mean before anything else moves.** Make these specific, measurable, and time-bound. How much partner-influenced pipeline in the next 12 months? Which segments? Which solution areas? Who owns what? When both organizations agree on the definition of success before the work starts, everything that follows has a north star.

2. **Build a progression model, not a binary one.** Darryl's zero-to-25 framework captures how partnerships actually build. Each one has a curve. Rather than measuring whether or not you hit a certain number,

track where you are on the curve and make deliberate moves to advance.

3. **Measure influence, not just attribution.** The value you create even when you didn't source the deal is the hardest and most important thing to capture. If you only measure what you can attribute, you are underestimating the value your ecosystem creates, and most likely underinvesting as a result.

4. **Make the invisible pipeline visible.** Deals are forming before you see them, in the moments leading up to the announcement. As a result, you may be missing a significant portion of your addressable market. Success in the decade ahead depends on account mapping, secure data sharing, and co-sell orchestration.

5. **Connect outcomes to customer success—always.** The most powerful proof point is not your pipeline number, but your customer's result. When the implementation works and the customer achieves what they set out to achieve, this evidence of a delivered outcome unlocks the next deal.

As Jen Odess of ServiceNow said, "We are almost the insurance policy of all global partnerships. Our job is to ensure the partner ecosystem is vibrant and AI-led from the day a partner onboards, to the day they are thriving and growing."

Customer success stories arm hyperscaler sellers with evidence. They attract new partners. They build credibility no marketing budget can buy. Deliver the outcome, demonstrate it, and the flywheel becomes self-sustaining.

Acres of Diamonds

Back in the 1870s, Russell Conwell, the founder of Temple University, delivered a speech that contained what is known as the Acres of Diamonds parable. It goes like this.

Desperately searching for wealth, a farmer sells his land to travel the world. He dies broke and empty-handed. Ironically, the new owner of his farm discovers one of the world's most lucrative diamond mines buried right under the property.

The idea is simple: the riches you are searching for usually are not found somewhere else. They're right there under your feet all along.

I have been drawing on this analogy since our earliest Ultimate Partner LIVE event. I also have a personal connection to it, as my daughter graduated from Temple University. Professionally, I feel it speaks to the incredible partner opportunities that so many organizations miss.

Kevin Peesker, then President of Microsoft SMC, delivered the keynote at our first UP LIVE event. He described a market of 38,000 accounts in the mid-market and commercial segments that were only 40% penetrated by partners. This meant the people gathered in that room were walking past tens of billions in annual revenue because they were too busy chasing the few crowded enterprise logos at the top of the pyramid.

His advice? Stop fighting over accounts already covered by global system integrators with armies of resources, and look around at the smaller players.

I said it from the stage that same day and it has been in my vocabulary ever since. "The SMC market is your acre of diamonds." So is the customer you already have won but have never fully expanded. So is the co-sell relationship that you've registered but not activated. So is the marketplace offer that you've listed but never enabled.

At the SMB level, there are an estimated 150 million customers worldwide that Microsoft will never touch directly. Every one of them represents an opportunity for a partner willing to show up. And when Microsoft directed the overwhelming majority of its channel incentives into the mid-market and SMB segment, the message was unmistakable. That is where the money went. And where Microsoft puts its channel incentives, partner revenue follows.

The greatest opportunities are rarely in the next campaign, the next product, or the next geography. They are in the ecosystem you have already built. The relationships you already have. The customers already working with you. The field sellers who already know your name.

Delivering outcomes, at its core, means mining the value already in your orbit—systematically, deliberately, and with the discipline to follow through. The infrastructure exists to do so now at a scale unimaginable five years ago. The marketplace motion, the co-sell automation, the partner data platforms, and the opportunity score are not just tools. They are the architecture of an outcomes engine.

Build it deliberately. Run it consistently. And become the partner everyone wants to work with, co-sell with, and build with.

YOUR TURN

Outcomes are the currency of trust. Everything you have built—the mindset, the commitment, the vision, the focus, the brand—can be validated only by what you deliver. These questions move you from an outcomes mindset to an outcomes system.

1. *Consider Darryl Peek's progression from zero to 25 co-sell deals. Where are you in that progression with your most important hyperscaler? What is the specific next milestone, and what would it take to reach it?*

2. *If your most important hyperscaler pulled your numbers today—your opportunity score, your co-sell record, your influence revenue—would they tell the story you want told? What is the one piece of proof you cannot currently show?*

3. *Think about the acre of diamonds in your current ecosystem. What is the biggest untapped opportunity already in front of you: a customer not yet expanded, a co-sell relationship not yet activated, a marketplace offer not yet listed or not yet enabled for self-service?*

YOUR COMMITMENT THIS WEEK:

Pull your co-sell pipeline and identify the three deals that have been stuck the longest. For each one, name the specific blocker and the specific action that would move it forward. Take that action this week.

CHAPTER 10

PRINCIPLE 7:
AGILITY

Not long before he died, legendary business thought leader Clayton Christensen stood on stage in front of Microsoft's most senior leaders and told us we were on a path to failure. But we didn't know it.

The occasion was PRISM, the PRIority Setting Meeting. Every year, Microsoft gathered its top executives in Redmond for a deep dive into the business. The highlight was always a session with an expert whose ideas spoke directly to the challenge of the moment. Past speakers had included Jack Welch, the CEO of GE, and *7 Habits of Highly Effective People* author Stephen Covey.

I wasn't sure why our new CEO Satya Nadella had picked this Harvard Business School professor to speak to us, even though

Clayton Christensen's argument had made *The Innovator's Dilemma* famous.

Well-managed companies doing everything right by the standards of their current business model can still fail, Christensen wrote. Not because they are incompetent, but because they are too committed to what already works. Existing customers demand improvements to existing products. Resources flow to the highest-margin opportunities. The organization optimizes for what is known. Meanwhile, a newer, leaner company sees what's needed in the emerging reality and captures the market. This is the essence of disruptive innovation.

Consider what was happening for us Microsoft executives at that moment. The organization had spent three decades building one of the most successful companies in history based on a model it understood perfectly: software licenses, enterprise agreements, reliable margins. Now the cloud was demanding that we disrupt ourselves… or watch someone else do it for us.

The world's leading expert on innovation was not presenting us with a theory. He was telling us to strike a new course or risk failure. Our leaders had gathered to set the company's priorities for the year, and Christensen was explaining how the priorities that had made us successful were the problem.

After the event, I understood Satya's choice. He was all about adopting a growth mindset. Christensen was telling us that we had to build agility into our business model if we wanted to survive.

The parallels to the partner ecosystem are direct, and I have watched failures in agility play out more times than I can count. A partner masters the Microsoft direct sales motion, then gets disrupted by the marketplace. A reseller builds a profitable business on perpetual licenses, then finds the market has moved to consumption or outcomes. A services firm becomes the expert in a single hyperscaler's architecture, then wakes up to find AI has rewritten the requirements. In each case, the organization was succeeding while the market was quietly moving past the motion.

That is exactly what makes this principle so challenging and so consequential: the moment agility matters most is the moment when your current results are telling you everything is fine.

Four Practices for Optimizing Agility

You have done everything right.

You did the internal work first. Your organization's mindset is on point, and from C-suite to field sellers, your people genuinely believe that partnerships pay. You have real executive commitment—funded and structured, not simply announced. Your vision is solid. Then you did the external work. You locked arms with organizations who see the same future you do. You're applying maniacal focus and delivering outcomes, ensuring your brand carries your story into the room before you arrive.

You may even have been named Partner of the Year.

And then your hyperscaler alters its priorities. The world changes. ChatGPT launches in November 2022, and within months every conversation, every budget, and every buying decision in the industry bends around AI.

Now what?

The answer to that question is why I have come to believe that the seventh principle is the most consequential of them all. The first six principles build the machine. Agility keeps the machine alive when the ground moves.

The best organizations I know are never surprised when the earthquakes hit, inevitably and with increasing frequency, because they are engaging in four practices the whole time. As a result, they pay attention to the market so that they can predict what comes next, and stand in a position to act before necessity forces their hand.

So assess your position, plan against the hyperscaler calendar, build an organization that can pivot, and stay technically current. Implementing these four practices ensures your organization remains agile, able to thrive in the midst of a constantly shifting landscape.

CHAPTER 10

Practice One: Assess Your Position

You cannot respond if you don't know what's going on. This first practice is a way to develop structural sight.

Conduct a quarterly SWOT assessment of your hyperscaler relationships and your ecosystem position. Every 90 days, ask the same four questions: What are we doing better than anyone else? Where are we underperforming? What new opportunities have emerged in the last quarter? What new threats, competitive, programmatic, or organizational, require a response?

Now pair the assessment with data. As Global Vice President of Partner Sales at Elastic Alyssa Fitzpatrick put it, become your own data scientist. "It's no longer one-year, two-year roadmaps, but three months. And by the fourth month, something could pop that you didn't know about." Watch your pipeline, your co-sell conversion, and your program economics, so that you can learn to see the correction coming before it arrives.

One organization I advised ran this exercise last fall and discovered a threat they would not have surfaced otherwise. Every new Microsoft incentive that quarter pointed at AI workloads, but none of their proof points did. They rebuilt their case studies as a result of their assessment, and resolved the issue in 90 days, before it was a problem.

Practice Two:
Plan Against Their Calendar, Not Yours

The hyperscalers reset their fiscal priorities every year. They restructure their field organizations, identify new solution areas, and reweight incentives. Most of these changes are not surprises. They are scheduled in advance.

Yet most partners experience the first day of the hyperscaler's new fiscal year as a crisis. They find themselves scrambling to re-qualify for programs that feel like they changed overnight—even though they didn't.

The directive here is to run your planning cycle ahead of theirs. Track the signals in the quarters before the reset. Invest ahead of the change, so you arrive on day one already aligned to the hyperscaler's new priorities, with your assets and proof points ready.

Tony Safoian built SADA's entire business on this philosophy. He called it a "culture of anticipation." On the first day of a new hyperscaler fiscal year, his team was already positioned for what the next year would require.

"You have to be able to invest in a controlled fashion in areas that you think, or know, are tremendous opportunities that may be too early for anyone to have recognized," Tony said. "You skate not to where the puck is, but to where the puck is going."

That's planning ahead.

CHAPTER 10

Practice Three: Build an Organization That Can Pivot

Anticipation is wasted if your organization doesn't move. You have to be structured so that you can pivot when you need to pivot.

Lori Borg, former Vice President of Go-to-Market for Microsoft Americas, calls this organizational agility. "When you're tied to a Microsoft funding program and the funding dries up, you need to quickly go do something different. Or Microsoft rolls out a new solution area or new competency requirements, you need to be able to pivot."

Audit your organizational agility by asking these questions: How much of your partner economics depend on funding programs you do not control? How long would it take to redirect your best people to a new solution area? Do your compensation plans allow a mid-year change of motion, or do they lock your field into last year's strategy?

Perhaps the most important question of all is: who has the authority to call the pivot? As Lori stated, this directive really must come from the top down. "It impacts how you invest, the horizons that you're planning things on, the rigor and discipline that you put around processes, so that you can move quickly if you need to."

SHI International, one of Microsoft's largest resellers, serves as an example of how to put agility into practice at scale. Built as a transactional powerhouse, they restructured for the services era by developing one operating system across their Microsoft practice. Pre-sales and delivery fell under one rhythm, and their go-to-market mirrored Microsoft's so closely that when the solution plays changed—as they do every six to 12 months—SHI's map changed with them.

Practice Four: Stay Technically Current

If you are not using the technology yourself, you cannot credibly lead customers through a transformation. This is technical agility.

Before your next customer pitch, ask what tech capabilities your organization has built and deployed. In today's world, this means using AI to accelerate your co-sell motion, surface buying signals, write better proposals, and train your field. Then you can use your internal transformation as the proof point you take to market. SHI did this when it deployed Copilot inside before taking it to Microsoft clients.

This practice is, in essence, an implementation of the frontier mindset we talked about in Chapter 4. You become your own customer zero. Move toward the edge before the crowd arrives, while the organizations watching from the sidelines fall behind.

CHAPTER 10

The Measure: AQ

At an Ultimate Partner LIVE event, Regina Manfredi, then Executive Vice President and General Manager of Crayon US and now President of North America at SoftwareOne, introduced us to a new term I've adopted ever since. We all know about IQ, or intelligence, and we've spent twenty years learning that EQ, or emotional intelligence, matters just as much. The new era, she said, demands a high AQ, or Adaptability Quotient.

IQ tells you how smart you are. EQ tells you how well you work with others. AQ tells you whether you will still be relevant in two years' time.

Regina's framing of the issue landed the way the best ideas do. Not with a gasp of surprise, but with a sigh of recognition. Every leader in that room knew a partner that was thriving five years ago, but whose strategy no longer applied. We all could name an organization that mastered a hyperscaler program, only to find it restructured overnight. We were all friends with a Channel Chief who had built a machine around a motion the market had already moved past.

While Regina calls it the Adaptability Quotient, I prefer to name this principle Agility. In my view, adaptability means possessing the capacity to change, whereas agility puts that capacity in motion. AQ measures whether you can read a shift.

Agility measures what you do about it. Build the first so you can practice the second.

This language is not mine alone. At a recent Google Cloud partner event, I heard leaders speak about "agility" from the stage a dozen times. At one of our own events, AWS marketplace leader Matt Yanchyshyn opened his advice to partners by borrowing the seventh principle straight from our slide.

Hyperscalers are saying it out loud, and they are watching for it in their partners. Agility is a must.

The Permanent Operating Requirement

Earlier, I posed the question: *now what?* You build for one reality, then it shifts. How do you handle the constant changes to hyperscaler priorities and market demands?

Here is the honest answer. There is no version of this industry where the questions stop arriving. In Chapter 1, when I spoke about the six shifts we're facing right now, I told you, "There is no after." We all live in a constantly and rapidly evolving world. The only response is to develop your AQ.

I have watched strong companies struggle to implement this principle. While they had great talent and solid resources, they fell behind when they stopped treating agility as an active practice and started treating it as a past achievement.

Agility is not a contingency plan you reach for when things go wrong. It is a permanent operating requirement, built into your planning cycles, your compensation structures, your product roadmaps, and your partnership operating model. When you do that, you can hold your position when the earth quakes, and keep your light burning when everyone else goes dark.

This may be the most consequential principle, because agility will keep your organization alive through every market shift, fiscal reset, and AI-driven transformation yet to come. Both the Internal and Partner Victories remain sustainable only when agility is built into your operating model.

The principles endure. Your tactics must evolve. That is the practice.

YOUR TURN

Agility is not reactive. It comes from running four practices on a rhythm: assessing your position, planning against the hyperscaler calendar, building an organization that can pivot, and staying technically current. These questions help you assess yours.

1. *Regina Manfredi talks about AQ, which tells you whether you will still be relevant in two years. On a scale of one to ten, what is your organization's AQ when it comes to partnerships today? What evidence supports your answer?*
2. *Which of the four practices is weakest in your organization today: assessing your position with data, planning against the hyperscaler's calendar, being structured to pivot, or staying technically current starting with yourself? What would it take to improve that practice this quarter?*
3. *Think back to November 2022. How long did it take your organization to respond to the launch of ChatGPT with an actual change in strategy, offerings, or investment? Whatever that number is, it is your baseline. The next shift is already forming. What would it require to cut your response time in half?*

YOUR COMMITMENT THIS WEEK:

Block one hour with your leadership team and ask one question: what is the most important shift happening in our hyperscaler ecosystem right now that we have not yet fully responded to? Name it. Then figure out the first action you will take to address it.

CHAPTER

11

ECOSYSTEMS:

THE NEXT FRONTIER OF GROWTH

The modern technology ecosystem was born just miles from where I am writing this book.

Decades ago, IBM ran a skunkworks team out of Boca Raton, Florida. In 1981, they made a decision that would change everything—though they did not fully understand it at the time. Rather than build everything in-house, they looked externally for an operating system to power the new personal computer they'd developed.

They found a young Bill Gates at a startup called Microsoft.

Gates did not have the OS IBM wanted. But he and his team quickly acquired one, purchasing a system called QDOS from a Seattle programmer named Tim Paterson. For around $50,000, Microsoft rebranded it as MS-DOS and brought it to IBM.

What happened next is one of the most consequential business negotiations in the history of technology.

IBM wanted to buy the operating system outright. Gates insisted on licensing it instead. At the time, IBM did not see the difference as significant. Operating systems were commodity infrastructure—the unremarkable plumbing beneath the real product, which was the hardware. Gates saw it differently. Microsoft would supply the OS to IBM, but retain the rights to license it to anyone else building IBM-compatible hardware.

That single decision transformed everything. As IBM licensed the PC architecture to clone manufacturers including Compaq, HP, and dozens of others, the market for IBM-compatible machines exploded. And with every machine sold, Microsoft's operating system went with it. Licensed repeatedly, at scale, across an entire industry.

The result was not just a successful product. It was the birth of a platform. And with it, the birth of the modern technology ecosystem.

Developers wrote software once and deployed it across a universe of compatible machines. Hardware manufacturers built to a certain standard. A new class of partnership emerged around the PC revolution. Resellers, developers, system integrators, and service providers all began orbiting the same platform. Everything depended on its success, and contributed to its growth.

At Itronix, we were an OEM—one of the thousands of companies creating value on top of the Microsoft platform. We took that very operating system I just described and repurposed it for the federal government. That was my seat at the table.

Then Microsoft recruited me. When they brought me aboard, they asked me to join their OEM offsite and give the room my perspective as a customer and partner of the company I had just joined. The platform owner wanted its newest insider to explain what the ecosystem looked like from the outside.

I never forgot that external point of view. As I ran partner businesses inside Microsoft, I made sure every program I built treated our partners as the vital participants in our company's success that they were. The best partner leaders never forgot which side of the table I came from.

Everything we now call "an ecosystem" traces back to that moment in 1981 when Bill Gates created the Microsoft licensing model. Michael Dell started building PCs in his dorm room. Compaq became the fastest company to reach a billion dollars in revenue at the time. Resellers like CDW turned direct marketing into an industry.

But here is the part most people miss. The ecosystem did not just happen *to* Microsoft. Microsoft recognized what it had built and nurtured it deliberately: organizing partners into specializations, standing up an entire partner organization, and aligning its salespeople to ecosystem success instead of siloing them away from it. No company has done it better. It is the blueprint every hyperscaler has been following ever since.

And it all started down the road from Ultimate Partner headquarters in Jupiter, Florida. The IBM campus where they built that first PC is an office park now. You can drive past it and never know that the modern technology economy was born inside those buildings. There is no monument. The monument is the industry itself.

The Channel We Grew Up In

That model had a precise architecture, and for four decades it defined how technology reached the world.

The vendor built the product. The distributor bought it in volume and financed the channel. The reseller bought from the distributor, added a margin, and sold to the customer. Every company in the chain had a role defined by its position in the transaction, and every company made its money the same way: on the movement of the product. Distributor points. Reseller margin. Deal registration. Rebates. Market development funds. The names themselves told you about the economics. A VAR was a value-added reseller. The value lived in the resale.

As a leader at a hyperscaler, I recruited the resellers, negotiated the margins, and fought for the deal registrations. It was a good model. It financed the PC revolution, put technology into every business in America, and built thousands of partner companies, some of which grew into the tech giants that anchor the ecosystem to this day.

And then the cloud quietly broke its central assumption.

When software became a subscription running in someone else's data center, ownership stopped changing hands. The platform billed the customer directly. The value stopped living in the transaction—the very thing the entire chain had been built to move, mark up, and monetize.

As Microsoft migrated to the cloud, our transactional partners were not happy, and they told us so. The margins they had built entire businesses on were evaporating, and from where they sat, we were the ones taking them away. So I traveled the circuit, partner event after partner event, making the case for changing their models to align their business models with what we were building.

The value in partnering with us was moving, I explained, but not disappearing. They could still generate revenue, mostly in new streams built around customer outcome rather than product. This included pre-sale services, post-sale services, deployment, adoption, and managed services. What I was describing, though none of us had the language for it yet, were the seats at the customer's table. The roles that would come to define the modern ecosystem.

Some partners heard me and adapted. Some did not. You can trace both groups forward to today. The ones who changed their models are the ones still growing. The ones who didn't are gone.

The ecosystem that surrounds hyperscalers today is not the old channel with new logos. It is a different organism entirely. The structure changed. The chain became a web.

The Hyperscaler Models

None of the three hyperscalers—Microsoft, Amazon, Google—started as a cloud company. Each arrived through a different door. Where they came from explains how they partner today.

Amazon began with infrastructure. In the early 2000s, Amazon realized their engineers were losing months on every project, rebuilding the same infrastructure before the real work could begin. Andy Jassy, then Jeff Bezos's technical assistant, saw the pattern and drafted a vision paper: turn Amazon's infrastructure itself into a product anyone could rent.

Bezos made the bet. When AWS launched, charging for server time by the hour on a credit card, the retail DNA was visible immediately: easy to buy, easy to start, easy to scale. That instinct produced the first cloud marketplace in 2012. The transaction infrastructure transformed how enterprise software is discovered, purchased, and deployed.

Microsoft's path ran through the PC revolution. Building upon the IBM licensing deal, the company spent four decades as a software and platform business with the most extensive partner ecosystem in the history of technology.

The cloud transition we made under Satya Nadella was not a pivot away from that ecosystem, but rather a reimagining of it. Instead of partners building on Windows, they started to build on Azure. Instead of reselling licenses, they could co-sell cloud solutions. The partner program that powered Microsoft's growth since the 1980s became the foundation for a full-stack model built around co-creation, co-selling, and joint customer outcomes—now accelerated by AI.

Google's pivot came from data. It had built extraordinary infrastructure to power Search, YouTube, and Gmail, arguably more advanced products than anything else in the market. Google Cloud was the externalization of that internal advantage. Which is why its differentiation has always leaned hardest into AI, analytics, and open-source tooling.

Three origins. Three instincts. One conclusion: the platform wins when the ecosystem thrives. AWS built the marketplace infrastructure first. Microsoft brought the deepest partner network. Google brought the most advanced AI and data capabilities. Together they created the hyperscaler era, and the ecosystem around them is now the primary engine of enterprise technology growth worldwide.

Few people have seen all three companies from the inside. But Darren Mowry has. He joined Google Cloud after a decade at Microsoft and ten years at AWS. When I asked him to compare the three as ecosystem partners, he declined the premise.

"I don't spend a lot of brain cells comparing what we're doing here at Google to what I experienced at AWS and Microsoft," Darren said. "What the market is telling me, and what customers are telling me, is what matters. The partners that are winning are the ones who start there, with the customer, rather than with a comparison of programs."

That customer-first orientation shows up consistently. Chelsea Berlucchi, who led co-sell for Google Cloud North America, gave me advice I have used in advisory conversations ever since: lead with a five-minute pitch and a five-minute win story, not a 40-slide deck.

"Get organized with the basics and the foundation there," said Chelsea. "How and why would a customer adopt your solution, your service, your technology? Those things will go a long way."

Simplicity is the point. Hyperscaler sellers manage enormous portfolios and enormous quotas. The partners who make themselves easy to understand, easy to position, and easy to trust are the ones who get called first.

Seven Seats at the Table

In Chapter 9, I described Jay McBain's 28 Moments analysis. It shows when enterprise deals are actually won. Not when the announcement is made, but months earlier, usually in paid engagements that most vendors never see.

The logical follow-on question is who, in a five trillion dollar industry, wins these engagements early on?

We have talked about this issue at our Ultimate Partner LIVE events for years. The buying decision now belongs to the empowered customer who assembles a team and decides which vendors to select for each aspect of the solution. The era of the single trusted advisor who could do it all is gone. The customer is not looking for a generalist. They are looking for the right specialist at the right moment.

Jay's data from Canalys, which has now been integrated with Omdia, is consistent. The average enterprise technology purchase is surrounded by 6.3 partners. Let's call it seven. Seven distinct partner organizations play a role in the customer's journey, including system integrators, ISVs, consultants, managed service providers, resellers, technology advisors, and more.

Jay is direct about the implication. "We are not selling commodities," he said. "We are selling highly considered purchases to a new, younger buyer who surrounds themselves with seven trusted people. The permutations for most vendors reach into the tens or hundreds of thousands of people who surround the buyer in their total addressable market."

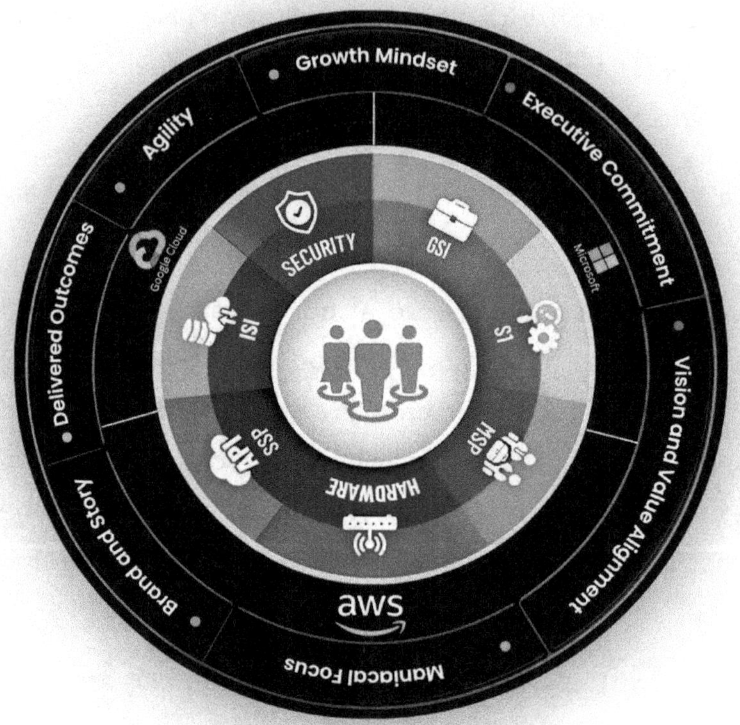

Figure 1. Seven Seats at the table, empowering the customer through hyperscaler platforms, surrounded and upheld by the Seven Principles.

In the AstraZeneca deal I pointed to in Chapter 9, Jay McBain analyzed how SoftwareOne and NTT won five seats at the table before other vendors even knew what was happening.

That is what the Seven Seats mean in practice. They are not won through a campaign or co-sell registration. They are won through years of relationship building, consistent delivery, and the kind of trust that only comes from showing up when it is inconvenient.

Rodney Clark, who served as SVP of Partnerships at Cisco, corroborated from the vendor side that mid-size organizations typically have seven or more partner companies engaged in sustaining outcomes. Large enterprises can have far more, 20 to 25 even. Cisco's partner program has been adapting to that reality.

"Our program has to evolve so that we're making those logical connections," Rodney said. "Creating a scenario where we're basically serving up those seven entities that are going to deliver outcomes, or we're facilitating the connection of more. That's a really significant shift from years past."

Every partner organization needs to answer this question honestly: In your target market, for your target customer profile, do you hold one of the Seven Seats? If not, what would it take to do so?

Fewer Partners, More Revenue

Greg Sarafin, formerly a Global Manager of the EY Partner Ecosystem, gave this book one of its most counterintuitive lessons back in Chapter 6, when I was addressing vision and value alignment. He built EY's alliance business from under a billion dollars to more than $10 billion on a deliberately constrained set of partners. Fewer strategic partners, he showed us, equals more revenue. Spread the same investment across hundreds of partners, and you'll get thin relationships everywhere, but deep relationships nowhere.

Here is what that conviction means in terms of how you structure your own ecosystem.

The instinct in most organizations is to grow the partner count. The thinking goes, "More coverage means more reach, which means more logos in the program." But winners do the opposite. They make hard choices about which few relationships deserve genuine co-innovation and co-investment, building their short list with the same rigor they apply to their most important customer accounts. And they treat everything else as platform participation.

From his vantage point at EY, Greg described how the deepest relationships lead to an orchestrator role. "Our clients expect us to help them navigate their most difficult journeys and to orchestrate a set of partners to help them do that. Not to say 'good luck working with Microsoft on your own!'" he explained. "We bring it all together. We translate. We take the client on a business outcome journey."

In the past, the alliance function carried a stigma. It was the place you sent people who could no longer sell. That's why Greg didn't want the job when EY first handed it to him. But then he changed the system by putting the customer's needs first.

This decade, we are dismantling the old negative perception of the partnership role entirely. In the modern ecosystem, the greatest value is being created by orchestration. And Greg gave me the best analogy I have ever heard for understanding what a platform actually does—and why the ecosystem around it can be so durable.

"You know what my favorite platform company is? RYOBI. Go to Home Depot. Those yellowish-greenish tools with the batteries? That is a platform. I have more yellow tools than anyone ever should. You're locked in. I'm on that platform."

I laughed because he was exactly right. Once you own the batteries, every new tool costs less than switching ever will, so you buy another then another, and the commitment compounds.

Microsoft, AWS, and Google Cloud are RYOBI at hyperscale. Their platforms create the same stickiness, lock-in, and compounding commitment. The partners who understand platform economics—who build their entire strategy around being indispensable within a platform—are the ones who will grow fastest and defend their position most effectively.

Finding Your Role in the Ecosystem

In the modern ecosystem, roles are not positions in a transaction chain. The old channel assigned your role by where you sat between the vendor and the customer. The new ecosystem assigns it by what you contribute to the outcome.

Four roles matter most in the modern ecosystem, and each one carries its own economics.

Innovators bring differentiated solutions to market, creating value that does not exist elsewhere in the stack. These are primarily ISVs: software companies building on top of hyperscaler platforms, embedding their technology into Azure, AWS, or Google Cloud and selling through the marketplace. Companies like AvePoint, Elastic, and ServiceNow have built enduring positions by going deep on a specific solution area rather than broad across many.

Innovators build recurring software revenue that grows with the platform. Every dollar transacted through marketplace retires customer committed spend, which makes the next dollar easier to win.

Enablers provide the tools, services, and platforms that help other partners and customers succeed. Distributors like Pax8, Ingram Micro, and TD SYNNEX make it easier for other partners to transact, enable, and scale. Co-sell platforms like PartnerTap and Tackle.io give the broader ecosystem the infrastructure to execute.

They earn on volume: small margins multiplied across the entire ecosystem's transactions.

Amplifiers extend the reach of others, scaling solutions across industries and markets that would otherwise be inaccessible. Technology solutions brokers, agents, and referral partners live here. So do the community organizations, media companies, and advisory firms like Ultimate Partner that help vendors and partners find and engage the right audience. Amplifiers may not transact directly, but their influence on buying decisions is real and measurable.

Amplifiers monetize trust itself via referral economics, audience, and influence over the moments where deals are shaped.

Integrators connect solutions into broader architectures, delivering the end-to-end value customers increasingly demand. Global system integrators like EY, Accenture, and Deloitte operate here, alongside regional SIs and the most sophisticated MSPs. They take pieces of the stack, such as the ISV solution, hyperscaler platform, and customer's existing infrastructure, and make them work together to deliver a business outcome.

Integrators capture the largest share of the multiplier: \$8 to \$11 of services that wrap every dollar of platform revenue, compounding through land-and-expand relationships that renew for decades.

The most successful organizations know their role and play it with deliberate focus. These days, they all are adding AI and Agentic as a layer of capability and as part of their playbook. Pick your role, and you have picked your revenue model.

The least successful organizations are trying to play all four roles simultaneously and end up being memorable for none. They're also collecting a chain-era margin while claiming an ecosystem-era role.

Knowing your role is not a limitation. It is the foundation of a defensible, scalable, compounding position.

YOUR TURN

It's time for you to think of the ecosystem not as a supporting cast but a primary growth engine. These questions will help you assess whether your organization is orchestrating the ecosystem or just participating in it.

1. *What is your role in the ecosystem: Innovator, Enabler, Amplifier, or Integrator? Is that role intentional and clearly communicated, or has it emerged by default? Is it the role you want to occupy for the next three years?*
2. *In a typical enterprise deal in your space, which of the Seven Seats are you occupying? Which seats are your competitors occupying that you are not? What would it take to earn one more seat at the table?*
3. *Greg Sarafin grew EY's partnership function by focusing relentlessly on fewer, deeper partnerships rather than many shallow ones. How many genuine co-innovation, co-investment partnerships does your organization have today—relationships where both sides are truly building together? Is that number too high, too low, or just right?*

YOUR COMMITMENT THIS WEEK:

Map your top five ecosystem partners against the four roles. For each one, identify whether the relationship is transactional, collaborative, or strategic. Then identify one relationship you want to move up a level, and schedule the conversation.

CHAPTER

12

THE NEW ROUTE TO MARKET:
MARKETPLACE, CO-SELL, AND GTM

Our industry never retires a route to market. Direct selling still exists. The channel still exists. Enterprise agreements and licensing programs still exist. What this industry does is create new routes on top of the old ones and move the growth to whichever one is fastest.

I know because I have lived nearly all of them.

In the early days of wireless, the route was a rep with a bag, and I carried one, selling directly to enterprises on our way to a public exit. At Itronix, the route was OEM: rugged computers built on Microsoft's operating system, moved through a defense channel I built. At Microsoft, the route was the enterprise agreement, and when it could not carry the cloud, my team crafted the first licensing model that could. Each of those routes was, in its moment, *the* route. Until it wasn't.

The displacement is happening again, right now, and faster than ever before. And I am building Ultimate Partner inside the newest routes: marketplace, co-sell and go-to-market.

The largest software companies in the world, which include CrowdStrike, Salesforce, Workday, and ServiceNow, have all moved to selling through marketplace. In fact, buyers purchasing through marketplace grew 400% over the last two years. Channel partners—companies built entirely on the traditional transactional model—have crossed a billion dollars in marketplace sales, faster than even they expected.

As Cyril Belikoff put it in February 2026, "Analysts predict that the Microsoft marketplace will be a $300 billion partner opportunity by 2030. We have literally seen our business doubling for three consecutive years." And channel-led sales have grown three to four times faster inside that doubling.

Unlike the previous displacements I lived through, this one is being driven by the buyer. In Chapter 1, I described how buyers have become smarter than sellers. The coming majority of technology buyers would rather complete a transaction through a cloud marketplace than sit through a sales cycle. The shift has arrived at the cash register.

Everywhere in our community, I see the same pattern. A partner leader arrives carrying three separate documents: a marketplace strategy, a co-sell program, and a go-to-market plan. Three owners. Three budgets. Three plans.

Put them away. These are not three separate disciplines, yet they are not one discipline either. They are integrated motions, each with its own distinct mechanics and rules, which only compound when they operate as one system. Your customer never experiences them separately. Neither should your organization.

Microsoft has used this mantra for years: "Build with, go to market with, sell with." When one of their go-to-market leaders stood on our stage and described the three motions "woven together like a braid," the image stuck for me. It is not enough to be great at selling with your hyperscaler, build solutions that land in the marketplace, or market together. The magic happens when you run all three as one.

Thinking clearly about that one system comes down to three questions. Where does the money live? How does it move? And how do you earn it?

One more thing before we dive in: let me state clearly that nothing in this chapter stands on its own. You will succeed if you have done the work of the first 11 chapters—if you've developed the trust, growth mindset, commitment, focus, vision, brand, track record, and agility that is required. If not, this is where your weaknesses get exposed.

Where the Money Lives

Stop hunting for a new budget.

Most partner organizations are doing exactly that: running demand generation, chasing net-new line items, fighting procurement for dollars that do not yet exist.

Meanwhile, the money is already there. Your customer has almost certainly signed a multi-year committed spend agreement with a hyperscaler. That capital is contractually locked—negotiated, approved, and waiting to be consumed. All you need to do is position your solution to help the customer retire a commitment they have already made.

Show up where the money lives.

Josh Greene, who spent five and a half years at AWS leading marketplace development before moving to Okta, calls this "gravity." Committed cloud spend has gravitational pull. It moves toward the partners who make it easiest to consume.

MongoDB and Snowflake, companies whose products compete directly with AWS in certain markets, chose to lean into AWS Marketplace anyway. That's not partnership sentiment. It's physics.

How the Money Moves

The plumbing underneath this industry has been rebuilt, economically and structurally. The economics are settled, but the structure is still taking shape.

Economically, marketplace fees collapsed to roughly 3% across all three hyperscalers. This is more a credit card fee than a margin, as Jay McBain put it. "It's much like a Mastercard or Visa swipe in consumer."

But the margin did not disappear. It just migrated to the work around the transaction: private offers, managed services, implementation, consulting, customer success.

"A partner might be willing to give up the 3% because that's a part of the business they're not all that interested in," Jay observed. "And they might be very interested in the other 17%." Repositioning your business around that gap is critical to becoming a successful partner today.

Structurally, things continue to evolve. Over the past year or two, every major hyperscaler has wired its channel directly into its marketplace. With Microsoft's Resale Enabled Offer, AWS's Channel Partner Private Offer, and Google Cloud's Marketplace Channel Private Offers, an ISV authorizes a channel partner to resell through marketplace. The channel partner assembles the offer, adds services, and brings it to the customer. The transaction runs through marketplace, retires the customer's cloud commitment, and the channel partner keeps the customer relationship.

Canalys projects that by 2027, more than half of all marketplace business will flow through the channel this way.

But projections are not the current reality, and I will not pretend the motion is settled. In conversations across our community, the same struggle comes up again and again: customers are asking whether they are supposed to buy from the hyperscaler or from their channel partner, and even the partners are unsure.

Here is the answer, and it is worth internalizing because the confusion is costing you deals: it is not either / or. The customer buys through the marketplace *and* from their trusted partner in the same transaction. The committed spend burns down, the hyperscaler counts consumption, and the partner owns the relationship and the services around it. The partners who are excelling right now are the ones who walk into the account and explain the model before anyone else does, because whoever explains the motion tends to own it.

And watch where the money enters the route, because this is the part most channel-era leaders miss. Matt Yanchyshyn, VP of AWS Marketplace, pointed out that well over a quarter of private offers begin with a self-service or product-led motion. A buyer finds the product, starts small on their own, proves it works, and then the deal graduates into a private offer that burns down committed spend.

The self-service start is no longer a small business play. It is how a new generation of buyers—CFOs and CTOs who grew up on their phones—begin complex enterprise procurement. They start without talking to anyone, researching in Claude and

ChatGPT, and then hand over a procurement decision already made. Partners without a self-service motion, in Matt's words, "are literally leaving money on the table."

This is not solely an AWS phenomenon. Google Cloud is investing in the same front door, surfacing partner solutions in context, inside the console where builders already work. The lesson is not that sales teams are obsolete. It is that the first mile of the route is now self-serve, and the deals that close biggest often start smallest.

Dai Vu, who leads marketplace and ISV go-to-market for Google Cloud, marked the moment the company moved marketplace from a preferred route to *the* route. "All the incentives, all the field compensation, all the time-to-value benefits are driving to marketplace," he said. His field data demonstrates faster deal cycles, higher win rates, and larger deal sizes, with multiple deals crossing $100 million in total contract value. The results now reach Alphabet's earnings calls. As I write this, CEO Sundar Pichai has reported marketplace transactions growing more than seven times year over year.

The early proof is encouraging. Cyril shared multi-party offer data on our stage. A third of marketplace sales flowing through the new motion within its first markets, at nearly double the average deal size. When Trade Ledger, an end-to-end lending platform, did their first co-sell deal with a major bank through marketplace, a deal previously negotiated at $4 million ended up closing at $25 million.

We have been here before.

In 2012, channel partners feared cloud meant customers buying direct and resellers getting cut out. I was inside Microsoft watching partners make that bet in both directions. The ones who embraced cloud found it created more services revenue, more consulting work, and more customer success opportunities than license transactions ever could. The ones who resisted plateaued. I remember who stood on which side of that bet.

The channel-through-marketplace motion is the next version of that story, still forming. The side you pick while it forms will matter just as much.

Beware, however, that a listing is not equivalent to a strategy. John Jahnke of Tackle has watched thousands of ISVs enter the motion. He said, "It's surprising to me that people still think, 'If I list, people just buy.' Like the field of dreams. I would love for that to be true. It's still not true. Listing is the starting point in the journey, not the destination."

How You Earn It

Now the issue I care about most. The money is committed. The rails are built. But that doesn't mean you're earning your share.

This is where partnerships are actually won and lost, and it is where I have spent my entire career. What I've found is that across all the shifts, how you earn has never changed. It happens one trusted relationship, one account, and one delivered outcome at a time. What has changed is the scale the motion now demands, and the discipline it rewards.

At Microsoft, back when the motion did not yet have a name, my team ran co-sell experiments. AWS had been building the marketplace version for years before most of the industry noticed. When I saw what Tackle was doing, I brought them on the podcast. I have been a believer in marketplace ever since.

Let's start with what co-selling actually is. Cassandra Gholston of PartnerTap drew a distinction that clarified my own thinking. "Co-sell is a transaction—one and done through marketplace. Co-selling is a verb. It's an action. It's top-of-funnel. How do we run plays? How do we orchestrate those plays before an opportunity is even created?"

Most organizations run co-sell reactively. A deal appears, someone registers it, someone waits for the hyperscaler to respond. By then the window has already narrowed. The partners I watch win run proactive motions are mapping the accounts they share with the hyperscaler, activating the white space, and orchestrating plays before any opportunity formally exists. They show up with evidence of work they started months prior to any deal getting signed. These are the acres of diamonds I mentioned in Chapter 9.

As Cassandra put it, "The technology is just the infrastructure. The companies that win at co-selling are the ones that have done the harder work, the human work, of building trust with their hyperscaler field teams." She is right. I watched an army of 50,000 sellers learn that lesson the hard way. No platform ever creates trust. People do.

Next, speak the seller's language. The field cannot bring you into deals it cannot understand. Pat Primavera of Microsoft said, "If your value proposition is not translated into Microsoft's language—if it doesn't connect to their solution areas, seller incentives, and fiscal priorities—you are invisible to the field. It doesn't matter how good your solution is."

It's also critical to understand what actually moves a field seller. Compensation drives behavior. Microsoft has wired marketplace directly into its sellers' pay. This is not just a partner-led motion, as Cyril explained. "Sellers follow their back pocket. That's literally how we ask them to operate."

Former SVP of Partnerships at Cisco Rodney Clark showed what happens when a vendor wires the whole system around compensation. He grew its hyperscaler co-sell business 40 to 55% year over year. "It's not an AWS and Cisco sell," he said. "It's an AWS, Cisco, and 'insert name of partner' sell."

You also need to know the price of admission. A transactable marketplace unlocks co-sell benefits across the hyperscalers, so any partner building IP needs a marketplace strategy wired into its go-to-market. Notice that the hyperscalers keep simplifying their own side of the motion, consolidating their solution plays so their sellers have fewer, clearer targets. You need to do the same. Your motion must tell a story simple enough that a seller can understand in 30 seconds why bringing you into a deal makes their number easier to hit.

At this point, it's critical you've determined what your One Thing is. Back in Chapter 6, I gave you three questions designed to help you hone in on the urgent problem you are uniquely positioned to solve, aligned with what the hyperscaler's leadership has publicly committed to, with clear revenue impact.

Here is its go-to-market consequence: when you commit to your One Thing, the whole motion becomes self-reinforcing. Your messaging resonates immediately with the customers who fit. Your co-sell registrations get attention because the solution area match is obvious. Your marketplace listing gets found because your keywords are specific. When you try to be relevant to everyone, you are compelling to no one.

Finally, run on their clock, not yours. As I discussed in Chapter 10, every hyperscaler operates on a fiscal rhythm, with quarterly priorities, annual solution areas, and funded programs announced at specific times of year. The partners who win consistently have reverse-engineered their plans according to that calendar. They show up with the right offer at the right time, not six weeks after the window has closed.

Becoming the Signal

Years ago, a Microsoft coworker of mine left the company—and left quite an impression. He titled his farewell post, "A

Shiny Quarter in a Bucket Full of Shiny Quarters." The image captured what it often feels like to compete for attention inside a hyperscaler ecosystem. You're up against tens of thousands of field sellers, carrying enormous quotas, and tens of thousands of partners all claiming to have the right solution for the right customer.

Here is the frame I'd like to offer instead: you are not trying to be the shiniest coin in the bucket. You are trying to be the lighthouse. The fixed signal that partner field sellers already know and trust, so that they turn toward you when they need help in your space.

You cannot become that signal by accident, and you cannot earn it through a single win wire email, portal update, or meeting at a partner conference. You earn it by showing up consistently. By understanding the seller's priorities better than any other partner in your solution area. By having the proof points, customer stories, and follow-through that make a seller's life better when they work with you.

You earn it by implementing the Seven Principles.

The Seven Principles in the Field

Here's what the Seven Principles look like on an operational level, at field altitude.

Executive commitment must show up everywhere. It is what drives your organization to hire and empower the partner-facing roles, GTM leads, partner managers, and solution architects who build genuine trust with hyperscaler field teams. Your sales, marketing, and customer success teams have to understand co-sell and be paid for it. If your comp plan works against the motion, nothing else matters.

Vision and values alignment brings precision. Microsoft, AWS, and Google Cloud organize their field teams differently, with distinct roles, incentives, and fiscal rhythms. If you treat every seller as generic, you'll misfire. You cannot run a shared motion with an organization you have not bothered to understand.

Maniacal focus sets your rhythm. You run QBRs, account planning, and regular syncs on the hyperscaler's cadence, so that you stay relevant through every fiscal transition. Brand and story become critical for sales enablement. Your messaging speaks the language of the seller's solution areas, not your product's features. Then, of course, you have to deliver outcomes. To do this, you need infrastructure: platforms that integrate with the hyperscalers' co-sell systems.

Growth mindset and agility keep your whole business alive. They ensure that you're piloting plays, learning fast, and re-mapping the motion every time the hyperscaler resets its priorities. Because it will.

And, of course, trust runs underneath it all.

The Seven Principles are not a pre-game exercise, but the architecture your organization runs on. And in the next chapter,

I'll show you the machine they assemble into: the flywheel, the payoff of all your hard work.

Blood to All Organs

In July 2026 at our AWS Executive Breakfast in New York, I moderated a panel about integrated motions. Matt Yanchyshyn of AWS Marketplace, Rekha Thangellapalli, who leads GSI Alliances at Elastic, Allison McFadden, who leads Accenture's North America AWS practice, and James Kang of NVIDIA, whose platform sits underneath the entire AI stack, all participated on the panel.

When AWS launched the capability at re:Invent that let channel partners bundle their services and data assets with an ISV solution on marketplace, Elastic and Accenture saw the opening immediately. Customers everywhere were chasing generative AI and hitting the same wall: their data was not ready. These two companies were already solving the problem outside the marketplace. They understood that the new mechanics gave them a way to package it as a repeatable offering, the Data Readiness Engine for GenAI. Each company had a clear reason to invest. Accenture got a differentiated, services-led offering. Elastic expanded its AI story. AWS drove marketplace adoption and cloud consumption. And the customer got a solution to a very real problem.

"The pitching, the idea, getting the budget, getting the executive sponsorship was actually the easy part," Rekha explained. "The hard part was getting all three companies to come together to go from idea to launch in a very ambitious timeline of six weeks."

So Elastic, Accenture and AWS built a RACI matrix and a project plan. They ran daily standups for six weeks, with at least one person from each of the three companies on every call, clearing blockers.

They launched on time. And the impact showed up fast. A global pharmaceutical company with nine petabytes of data spread across six continents used the offering to build the trusted foundation its AI agents could safely tap at scale.

This was an impressive feat. I have watched large alliances spend six weeks just scheduling the kickoff call.

"When you're designing any kind of offering or go-to-market motion, it has to get blood to all organs," Allison said. "If it does not get blood to all organs, it does not go."

I love that analogy, and I have been repeating it ever since. The customer is an organ. The ISV is an organ. The GSI is an organ. The hyperscaler's relationship, the marketing team landing leads on a storefront that converts, the field reps who retire quota when it transacts through marketplace—all are organs. The test for integrated motions is: does the offering feed every organ simultaneously? Starve one organ and the motion dies, no matter how good the technology is.

When it works, the returns compound beyond the transaction. Rekha's data at Elastic shows that bringing GSIs into deals increases renewal rates, expands deal sizes, and grows annual recurring revenue over time. The offering is the start. The compounding is the point.

The Agentic Era

One more displacement is already underway, this time inside the motion itself: the majority of partner opportunities will be progressed and managed by AI agents this year.

Matt Yanchyshyn said that the majority of customer opportunities at AWS already have a partner attached. The goal is for the majority of those attachments to come not from a human, but from the solution-matching engine.

How do you get recommended by an AI-driven engine? By building a healthy, well-maintained pipeline with clean data flowing through the integrations, and by developing specializations and multi-product solutions. Partners have already built agents on the Partner Central APIs managing more than 5,000 opportunities. The seller with 20 years of experience clicking opportunities through a pipeline is done. Those people belong in front of customers. The pipeline belongs to the machines.

What I'm describing is the frontier firm framework from Chapter 4, which Cyril Belikoff laid out. The matching engines will favor the organizations experimenting with agentic workflows inside their own walls. Mindset first, then motion.

I have watched this movie before. When cloud arrived, companies panicked over runaway costs and pulled back. Many of those companies no longer exist. The organizations winning now embraced cloud years ago.

The same choice sits in front of every partner today when it comes to agentic AI, on both the buyer and the seller side. Embrace it—put the governance in place, make your data available, move toward agentic co-sell—or fade and die.

I do not use that language lightly. I use it because I've watched the last set of holdouts disappear. The cost of experimentation has never been lower. The cost of waiting has never been higher.

So follow this directive, as plainly as I can give it to you:

Make your offer transactable, and sell into the committed spend your customers have already promised. That is marketplace.

Map the accounts you share with your hyperscaler, run on their clock, and deliver and log every outcome, so the sellers and the machines bring you the next deal. That is co-sell.

Build with, market with, and sell with your hyperscaler as one motion, behind a story a seller can repeat in 30 seconds. That is go-to-market.

And then give these directives force. Choose your role, because your role is your revenue model. Claim your seats, because a seat, once earned, pays you for years. Price yourself into the services and outcomes that wrap the platform, not the three points left in the transaction. And concentrate on the few motions that produce most of your value, refusing everything else.

The partners I have watched cross from millions to hundreds of millions are doing these few things at depth, for years, while everyone else samples everything at the surface.

Marc Monday runs the partner channel at ServiceNow. When I asked him what partners should do differently in today's world, he said, "Go faster." However fast you think you are going, it is not fast enough.

Then he laid out the arithmetic. His roadmap ships weekly, but the certification model this industry built takes 12 to 18 months. Take that long to finish it, and you are credentialed in a product that no longer exists. As Marc's CEO Bill McDermott put it, "The market has never been this fast, and it will never be this slow again." Just think about the pace of change with AI, and you'll understand what he means.

Marketplace, co-sell, and go-to-market are not three routes you need to get right once in order to succeed, because they will keep changing. The routes will continue to shift, again and again. The principles will not.

YOUR TURN

Marketplace, co-sell, and go-to-market are integrated motions—distinct disciplines that only compound as one system. Here are three questions I ask every partner leader.

1. *Where does the money live? Map your top ten customers: which hyperscaler holds their committed spend, how large is the commitment, and is your solution positioned to help retire it? If you do not know the answers, that is the first gap to close.*

2. *How does it move? Where is your organization on the maturity curve—no listing, listed but passive, actively co-selling, or the integrated motions running as your primary route to market? What is the single biggest thing preventing the next stage?*

3. *How do you earn it? Is your value proposition translated into the hyperscaler's language—their solution areas, their seller incentives, their fiscal priorities? Then find your weakest of the six essentials and name what it would take to fix it in 90 days.*

YOUR COMMITMENT THIS WEEK:

If you do not have a transactable marketplace offer, identify the single biggest obstacle and the specific person responsible for removing it. If you do, reach out to one hyperscaler seller you have not engaged in the last sixty days with something specific: a customer story, a proof point, or a solution play tied to their current priorities.

CHAPTER

13

THE FLYWHEEL

FULLY ASSEMBLED

I opened this book with a simple premise: Success is never built alone.

The Ultimate Partner is defined not by the size of their organization, the number of awards on their walls, their tier status within the Microsoft, AWS, or Google Cloud ecosystem, their marketplace revenue, or their co-sell attach rate.

The Ultimate Partner is defined by how they show up.

They show up with a growth mindset—genuinely open to change, genuinely committed to the abundance that partnerships create.

They earn executive commitment—not just at the top of their organization, but across every function, compensation plan, and performance metric.

They align on vision and values—with their hyperscaler partners, their customers, and the ecosystem organizations they co-create with.

They execute with maniacal focus—finding the One Thing that drives momentum and relentlessly pursuing it.

They build a brand and tell a story—making themselves easy for hyperscaler sellers to evangelize, easy for customers to trust, easy for the ecosystem to rally around.

They deliver outcomes—not as a destination but as a discipline, demonstrated and improved upon in every cycle.

And they stay agile—treating change as a permanent condition of the world they have chosen to operate in.

Now I want to close by bringing together the full picture with the flywheel. Because the Seven Principles only reveal their full power when you view them as a connected system rather than a list.

When you do everything right, fully embracing the Seven Principles, each engagement creates the next opportunity. Your efforts form a flywheel, and your momentum grows exponentially.

You are not grinding anymore. You are compounding.

Ultimate Partner LIVE

This is what the Ultimate Partner has figured out. Not how to execute one principle well, but how to operate all seven, simultaneously, as an integrated organizational capability.

I learned this through 40 years of experience on every side of the equation. And I see it happen, live, at every event we hold.

I have watched a partner leader arrive without a co-sell strategy and leave with a roadmap, a bench of peers who have already solved their exact problems, and relationships with hyperscaler leaders who now know them by name.

I have watched an ISV with a marketplace listing but no motion connect with another partner and leave with a multiparty offer that became material revenue within two quarters.

I have watched executives who talked past each other for months sit in one room, hear the same data, and align within two hours.

None of this happens by accident. It happens by design when you put the right leaders in the same room, ground them in the Seven Principles, and give them an opportunity to connect with other people who have already done the work.

That is what Ultimate Partner exists to create.

Four decades in the industry have taught me which organizations survive the shifts. Never the ones holding the best playbook for the current route. Playbooks expire. The survivors are the ones who build on the Seven Principles.

The route will change again. The principles will not.

The Work You Have Already Started

If you have been honest with the Your Turn exercises at the end of each chapter—and I hope you have, because if not, you're only cheating yourself—then you already know something important.

You know where you stand today.

You know which of the four failures is quietly undermining your strategy. You know whether your organization's growth mindset is a seven or a four. You know which of the co-sell essentials is your weakest link. You know whether your brand is making you the signal people navigate toward, or just more noise in an already crowded marketplace.

Gaining this level of insight into your organization is not comfortable. I know that. I have done this work on the inside at Microsoft, and on the outside as an advisor at Ultimate Partner. The gap between where you are and where you need to be can feel daunting when you first look at it clearly.

But I can assure you that the leaders who see the gaps are far ahead of the ones who have not looked. Most organizations never do honest reviews. They reach for a new tactic, a new program, a new hire—and wonder why the results do not change.

You have done something different. You have looked. Now the only question is what you do with what you saw.

My ask is simple. Do not try to fix everything at once. Go back to the chapter where your score was lowest. Find one commitment you wrote down. Do one thing this week. Then the next week, do the next one.

Results do not come from a great strategy alone. They come from the first action. Then the second. Then the third. And at some point—not overnight, but sooner than you think—momentum takes over, and your flywheel is in motion.

This is how every partner organization I have watched succeed has done it. Not with a grand transformation. With a series of deliberate actions, taken consistently, over time.

You have the Seven Principles. You have the framework. You have done the self-assessment.

The room is ready.

CONCLUSION

CHANGE
IS ACCELERATING

We're living in a time of continuous, momentous change.

Jay McBain declared the 2020s the Decade of the Ecosystem. We are almost seven years in at the time of this writing. The honest assessment is that the organizations who moved early are pulling away from those who are still debating whether to move at all.

The buyer is changing in ways that make the traditional channel model not just inefficient but structurally misaligned. By 2025, Millennials became the majority of global software buyers—a demographic that is, as Jay describes it, digital-first and digital-only. They grew up on subscription and consumption models. They prefer to buy through marketplaces rather than sit through sales cycles. They assemble Seven Seats now, and Jay has said publicly that as deal complexity continues to rise, that number could move toward 17.

In the not-too-distant future, some of those buyers will not be people at all. Jay pointed to the 2030s as the era when agentic AI begins making purchasing decisions on behalf of organizations.

Marketplace transformation has crossed a tipping point. Canalys projects that by 2027, more than half of all marketplace business will flow through the channel. The billion-dollar club has expanded from one to many. AWS has become a top five distributor by any measure. The organizations still treating marketplace as a side project will find their revenue streams increasingly difficult to sustain.

At the same time, the co-sell revolution is happening, but unevenly. The organizations that are winning have invested in people, systems, and cadence. Those trying to co-sell with spreadsheets and goodwill are failing.

And the AI transformation is accelerating everything. The pilot era is over. The accountability era has arrived. The partners who can deliver specific, measurable, customer-validated AI outcomes in specific industries, for specific problems are building an advantage that will compound for years.

The window is still open, but it is narrowing. The opportunity is massive. And it belongs to the organizations that do the work.

Finding Our Shared Purpose

When people ask me what drives this work, I give them the same answer I have given from stages, in studios, and in advisory conversations for nearly a decade now.

I am on a mission to help every leader, every organization, and every partner achieve their greatest results through successful partnering.

This is not a narrow mission. I do not speak only to the largest enterprise partners or the most sophisticated channel chiefs. I aim to include every partner at every organization, from the MSP building a managed services practice, to the ISV trying to get their first marketplace deal done, to the alliance leader inside a large company who cannot get their CRO to take partner revenue seriously, to the Channel Chief who believes in ecosystem-led growth but cannot get budget approved.

This book is for all of you. You are why the Ultimate Partner podcast exists, and why the LIVE events exist.

My hope is that the content we've created has served as a lighthouse, guiding you through some genuinely turbulent waters. And we are not done. The opportunity is too large, and the gap between where most organizations are and where they could be is still too wide.

I don't want to do it alone. You are a critical part of this effort.

The Ultimate Partners I have known across this journey share one defining trait. They achieved their greatest results in partnership: with their ecosystem, their hyperscalers, their customers, and their community.

Together, we can build the room. The right room. With the right people. At the right moment.

Guiding Lights

A month after my bike accident in the spring of 2025, I was lying in a hotel room at 2 AM, icing my leg before going on stage at our UP LIVE event the next morning. And I finally got it. The reason for my dedication to Ultimate Partner.

The first room that ever shaped me was the kitchen. Around the table, my hardworking grandparents and parents gathered with their offspring to cook and share meals. I watched them build businesses, family, and community on a foundation of trust. They did not come from much, but they had everything that mattered: core principles, a work ethic, and an unshakeable belief that their kids could do extraordinary things. Around that table, I learned that there are no shortcuts. The work itself matters. So do principles.

Nobody had to explain it to me then. Nobody has to now.

For 40 years, I have built on that foundation. I lived the Seven Principles for decades before I ever wrote them down. They are not theoretical. They are what works.

And the circle keeps closing. Over the past few years, two classmates from my childhood found their way back to me. We grew up at the same kitchen tables. They went on to build legacies of their own. One advises Ultimate Partner today, and his daughter works with us. The other keeps walking me into rooms I never expected to stand in.

Now, I'm passing the torch to the next generation. I am building Ultimate Partner with my son John by my side. The decisions we make together rest on that same foundation of trust and integrity.

The book, the podcast, the live events, the community—all are rooted here, around the same table. The table has just gotten bigger.

For years, the Statue of Liberty's torch burned bright over Ellis Island, inspiring immigrants who arrived in our country with little more than dreams of a better life. To this day, the lighthouse in Jupiter keeps burning for sailors navigating uncertain waters. Two lights. One hope. Both helping guide people home.

The room is waiting. And we're keeping the light burning brightly in it.

So if you are building something that matters, come build it with us. And bring along those who are coming up behind you, the ones who need to know that integrity beats every shortcut, and that relationships built on trust endure.

Will you join us?

Vince Menzione

Jupiter, Florida

2026

ACKNOWLEDGMENTS

This book would not exist without the people who showed up—on the podcast, at the UP LIVE events, in the advisory room, and in the moments when showing up was the hardest thing to do.

To my family. You are the foundation of everything. Through the long hours, the travel, the events built in 120 days, the bike accident that changed everything, and the recovery that tested us all, you never stopped believing.

To my wife, my Nurse Jackie, thank you for being by my side through every chapter of this journey, and for carrying the weight of so much so that I could keep building. You are the quiet force behind everything visible. Everything I do, I do because of you all.

To my son John, thank you for your strength and steadiness. You showed up when it mattered most.

To my daughter Allie, watching you build your life with grace and joy has reminded me what this work is ultimately for.

To my mother, you told me to "just do it" when I needed to hear it most. That phrase became a compass. You have been one of my greatest sources of inspiration, and I carry your voice with me every time I step onto a stage or into a room.

To my father, the values you instilled in me have been my most durable competitive advantage. In a world where integrity and trust are increasingly rare, you made them feel ordinary. They are anything but.

To my grandfather, I believe you are my greatest energy source. You came to this country with almost nothing and built something from sheer will, belief, and a refusal to accept limits. At this stage of my life, when I am building something that I hope will outlast me, it is your example I return to most. You inspire me to be and do more.

And to my grandmother, you showed me who I am. A child of immigrants. That identity—shaped by sacrifice and the belief that the next generation could have more—is the foundation beneath everything I have built.

To Coach T, Trevor McGregor. You were the one who told me to start the podcast. Ten years ago, when I was figuring out what came next after Microsoft, you saw something I hadn't fully seen in myself yet. You guided me toward this place. Without that conversation, none of what followed would exist. Thank you.

To the guests of the Ultimate Partner Podcast. Over 300 conversations, more than 400 leaders have given your time, talents, stories, and hard-won lessons to a community that needed them. You are the co-authors of this book in every meaningful sense. Without your voices, there is no framework.

Without your experiences, there are no proof points. Without your trust, there is no movement.

Special acknowledgment to the leaders whose voices appear most prominently in these pages. You trusted me with your stories. I have tried to honor them. Tony Safoian, whose partnership journey inspired the podcast and whose commitment to excellence has never wavered. Jay McBain, whose research and generosity of insight have shaped how an entire industry thinks. Dai Vu, Janet Schijns, Cyril Belikoff, Matt Yanchyshyn, Dux Raymond Sy, Satya Nadella, Rodney Clark, Kevin Peesker, Nicole Dezen, Nina Harding, Ashleigh Vogstad, Marc Monday, Greg Sarafin, Rebecca Jones, Alyssa Fitzpatrick, Reis Barrie, Erin Figer, Darryl Peek, Jen Odess, Cassandra Gholston, Lori Borg, Regina Manfredi, Josh Greene, John Jahnke, Pat Primavera, Rekha Thangellapalli, and Chelsea Berlucchi.

To the Ultimate Partner team. You have built something remarkable with me—from the studio in Boca Raton to the stages in Dallas, Redmond, Reston, San Francisco, Bellevue and NYC. Every event, every episode, every community interaction reflects your commitment and care. This book could not have happened without the infrastructure and belief you bring to this mission every day.

To our sponsors and partners. You placed your confidence in this work before it was fully formed. Your belief in the mission, and in the community of leaders we serve, made the LIVE events possible, the studio possible, and ultimately this book possible.

To the Ultimate Partner Community. You showed up. In Dallas when we were new and unproven. In Redmond when we had found our stride. In Reston, Bellevue and in every event since. You filled the rooms. You asked the hard questions. You shared what you were learning. And you reminded me, every single time, why being in the room changes everything.

To the Microsoft, Google Cloud, and AWS partner teams who have trusted this platform as a place to tell their stories, share their priorities, and engage authentically with the partner community. Your partnership with Ultimate Partner has given our community access to the conversations that matter most.

To Dr. Michael Gervais, whose work on high performance and the psychology of partnership has shaped how I think about every principle in this book. "Nobody does the extraordinary alone." I have never forgotten it.

And to you, the readers, who picked up this book because you believe in the power of partnerships, who are doing the hard work of building something that lasts, and who understand that the greatest results are never built alone.

And to anyone I have not named here—I am certain I have forgotten many, and for that, I am sorry. The community that has built around this work is larger than any acknowledgments page can hold. If you have been part of this journey in any way, you are part of this story.

PODCAST SOURCES
AND EPISODE REFERENCES

Every episode referenced in this book is available at ***theultimatepartner.com*** and on Apple Podcasts, Spotify, and all major podcast platforms. Our YouTube channel is: @ultimatepartner.

You'll find the full archive of more than 300 episodes there, with fresh content added every week.

Introduction

Episode 274 — Vince Menzione: Why Success Is Never Built Alone

Episodes 81/82 — Dr. Michael Gervais: How First Principles Guide and Impact Successful Partnerships

Chapter 1 — Your Strategy May Already Be Obsolete

Episode 290 — Nina Harding, Microsoft: The AI Pilot Era is Officially Dead

Episode 263 — Cyril Belikoff, Microsoft: The Lightbulb Moment: Microsoft's Vision for AI and Its Partners

Episode 223 — Dux Raymond Sy, AvePoint: How Data Management, Security and the Future of AI Leads to Success

Episodes 286/288 — Jay McBain: Why the AI Economy Is a Multiplier Game / The Millions You're Losing Without Even Knowing It

Episode 230 — Oguo Atuanya, former Microsoft GM (now Pax8): Decoding Partner Ecosystem and AI Innovation

Chapter 2 — Why Partnerships Fail: The Four Failures

Episode 147 — Vince Menzione: How You Recognize the Dysfunctions of Effective Partnerships

Episode 219 — Janet Schijns, JS Group: Why Partnerships Fail and How to Fix Them

Episode 260 — Jay McBain, Janet Schijns, Greg Sarafin: The Partnering Revolution

Chapter 3 — Trust

Episodes 81/82/168/250 — Dr. Michael Gervais

Episode 144 — Alvaro Celis, Microsoft: Applying Mindset, Just Cause and Agility to Achieve Your Greatest Results

Chapter 4 — Growth Mindset

Episodes 81/82/168/250 — Dr. Michael Gervais

Episode 263 — Cyril Belikoff, Microsoft: The Lightbulb Moment

Episode 270 — Nina Harding, Microsoft: Microsoft's Tectonic Shifts: A Partner Roadmap for FY26

Episode 211 — Jay McBain: Why Does Every CEO Need to Be a Partnership Ecosystem Leader?

Chapter 5 — Executive Commitment

Episode 214 — Greg Sarafin, EY: Unleashing $Billions in Growth: Chief Partner Officer Secrets

Episodes 64/126/210 — Tony Safoian, SADA

Episode 238 — Janet Schijns + Steven Pivnick: The New Era of Chief Partner Officer

Episode 267 — Alyssa Fitzpatrick, Elastic: The 1% Club

Episode 270 — Nina Harding, Microsoft: FY26 Partner Roadmap

Episode 211 — Jay McBain: Why Does Every CEO Need to Be a Partnership Ecosystem Leader?

Chapter 6 — Vision and Value Alignment

Episode 214 — Greg Sarafin, EY

Episodes 81/82/250 — Dr. Michael Gervais

Episode 291 — Rebecca Jones, Bridge Partners: The Power of Three

Episode 144 — Alvaro Celis, Microsoft

Chapter 7 — Maniacal Focus

Episode 229 — Erin Figer, CORE Consulting: 200% Pipeline Growth: Co-Selling Tips Revealed

Episode 267 — Alyssa Fitzpatrick, Elastic: The 1% Club

Episode 291 — Rebecca Jones, Bridge Partners: The Power of Three

Episode 275 — Reis Barrie, Carve Partners: 12 Costly Microsoft Partnership Mistakes

Episodes 214/233 — Greg Sarafin, EY

Chapter 8 — Brand and Story

Episodes 15/76/223 — Dux Raymond Sy, AvePoint

Episode 290 — Nina Harding, Microsoft

Episode 210 — Tony Safoian + Dee Burger: How Two Remarkable Tech Titans Redefine the Category

Chapter 9 — Delivered Outcomes

Episode 283 — Darryl Peek, Elastic: Hyperscaler Domination: How Elastic Won Pinnacle Partner Status

Episodes 284/286/288 — Jay McBain

Episode 229 — Erin Figer, CORE Consulting

Episode 303 — Matt Yanchyshyn, AWS: What's Coming — It's Tectonic

Chapter 10 — Agility

Episode 254 — Regina Manfredi: Mastering Tech Evolution: AQ, Bold Partnerships & Modern Selling

Episode 37 — Tony Safoian, SADA: Skating to Where the Puck is Going

Episode 266 — Lori Borg, Microsoft: Mastering Microsoft Partnerships: Adaptability, AI, and the Path to Growth

Episode 281 — Joseph Bellian, Stefanie Dunn (SHI) + Marcus Jewett (Microsoft): Why SHI's Audacious Transformation is Mastering Agentic AI

Episode 267 — Alyssa Fitzpatrick, Elastic

Episode 303 — Matt Yanchyshyn, AWS: What's Coming — It's Tectonic

Chapter 11 — Ecosystems

Episodes 282/284/288 — Jay McBain

Episode 214 — Greg Sarafin, EY

Episode 235 — Rodney Clark, former SVP Partnerships, Cisco: Decoding Customer-Centric Innovation

Episode 215 — Chelsea Berlucchi, Google Cloud: Co-Sell Secrets Exposed

Episodes 64/126 — Tony Safoian, SADA: Journey to Premier Status / Hyper-Focus as a Growth Imperative

Chapter 12 — The New Route to Market

Episode 192 — Julie Sanford, Microsoft: Game-Changing Moves at Inspire

Episodes 263/287 — Cyril Belikoff, Microsoft: The Lightbulb Moment / The $300B Marketplace Shift

Episodes 185/265 — John Jahnke, Tackle: The $200 Billion Opportunity in Cloud GTM / From First Click to $1B

Episode 279 — Erika Irby, Veeam + Pat Primavera, Microsoft

Episodes 303/304 — Matt Yanchyshyn, AWS: What's Coming — It's Tectonic

Episode 266 — Lori Borg, Microsoft

Episode 231 — Cassandra Gholston, PartnerTap: Unlocking Revenue Growth

Episode 80 — Natalie Singh, Microsoft: Authentic Leadership, Focus, and Grit

Episode 136 — Jake Swenson, Microsoft

Episode 259 — Jay McBain: Tech Trends and Predictions

Conclusion

Episode 274 — Vince Menzione: Why Success Is Never Built Alone

Episodes 82/250 — Dr. Michael Gervais

Episodes 251/282/284 — Jay McBain